Water Conservation and Environmental Stability

Water Conservation and Environmental Stability

Edited by **Keith Wheatley**

SYRAWOOD
PUBLISHING HOUSE

New York

Published by Syrawood Publishing House,
750 Third Avenue, 9th Floor,
New York, NY 10017, USA
www.syrawoodpublishinghouse.com

Water Conservation and Environmental Stability
Edited by Keith Wheatley

International Standard Book Number: 978-1-68286-035-9 (Hardback)

Printed in the United States of America.

Contents

Preface

Water conservation focuses on developing strategies and measures for sustainable utilization of water resources. This book covers the various topics that are contributing to the growth of the discipline such as hydropower generation, rainwater harvesting, modeling of groundwater resources, drinking water supply and distribution, etc. It contains case studies and researches by internationally acclaimed experts. This text is a vital tool for all researching and studying this field.

All of the data presented henceforth, was collaborated in the wake of recent advancements in the field. The aim of this book is to present the diversified developments from across the globe in a comprehensible manner. The opinions expressed in each chapter belong solely to the contributing authors. Their interpretations of the topics are the integral part of this book, which I have carefully compiled for a better understanding of the readers.

At the end, I would like to thank all those who dedicated their time and efforts for the successful completion of this book. I also wish to convey my gratitude towards my friends and family who supported me at every step.

                                                                                                    Editor

Measurement of the Cooling Efficiency of Pavement-watering as an Urban Heat Island Mitigation Technique

Martin A. Hendel[*1,2,3], *Morgane Colombert*[2], *Youssef Diab*[2], *Laurent Royon*[3]

[1]Paris City Hall, Water and Sanitation Department, F-75014, Paris, France
e-mail: martin.hendel@paris.fr
[2]Université Paris-Est, Lab'Urba, EA 3482, EIVP, F-75019, Paris, France
[3]Univ Paris Diderot, Sorbonne Paris Cité, MSC, UMR 7057, CNRS, F-75013, Paris, France

ABSTRACT

The Paris region (Île-de-France) was amongst the hardest hit by the August 2003 heat wave, due in part to subsequent amplification of its urban heat island. This has created high heat-wave awareness in climate change adaptation studies for the city of Paris. Over the summer of 2013, pavement watering was studied experimentally in two locations as a climate change adaptation method. Pavement watering was found to lower pavement surface temperatures by several degrees for several hours after watering, while also strongly reducing its cooling rate a few hours before and after sunset. Heat flux and storage at a depth of 5 cm in the pavement were also found to have been significantly reduced, especially during direct sunlight exposure, but also when the pavement was in the shade. Uninterrupted watering appears necessary during direct sunlight exposure of the pavement to maximize efficiency.

KEYWORDS

Urban heat island (UHI), Climate change adaptation, Pavement watering, Pavement heat storage, Pavement surface temperature.

INTRODUCTION

The Île-de-France region was one of the worst hit in Europe during the August 2003 heat wave, due in part to subsequent amplification of its Urban Heat Island (UHI), with a two-fold increase in mortality over the first two weeks of August 2003 [1]. This event created strong heat-wave awareness in France. As a result, increases in heat-wave events are at the heart of climate adaptation work for the city of Paris under its 2007 Climate Plan.

The city of Paris funded a research project that was undertaken from 2008 to 2012 [2]. Among other aims, this work simulated different adaptation methods on the basis of a Town Energy Balance (TEB) [3] simulation of the 2003 heat wave in Paris. Among methods tested, daytime city-wide pavement-watering was investigated. Following this study, a field experiment of pavement-watering was conducted during the summer of 2012 [4]. Results showed that apparent pavement surface temperatures were reduced by up to 7 °C.

Internationally, field studies have been conducted in many cities, particularly in Japan, on both traditional impervious pavements and porous or water-retaining pavements [4-8]. Yamagata *et al.*, watering a porous water-retaining road pavement with reclaimed waste

---

water in Tokyo, found surface temperature reductions of 3-8 °C [7]. Kinouchi and Kanda, using snow melting pipes to water an impervious pavement in Nagaoka City, found up to 10-30 °C of surface cooling [6]. They also investigate pavement heat flux and its links to net radiation. In Paris in 2012, Bouvier *et al.*, watering the sidewalk and pavement once at night with cleaning trucks, also found 3-8 °C in surface cooling [4]. Prior work from Asaeda *et al.* provides information on the behaviour of pavement heat flux under normal conditions in asphalt or concrete 20-30 cm thick slabs [9]. They also highlight the dominant role of net radiation.

These experiments provide valuable data for the surface temperature and reduction effects of pavement-watering. However, only limited attention is given to pavement heat flux and storage effects. Furthermore, data available for such trials in Western Europe are quite scarce. Our field study was conducted over the summer of 2013 in Paris, France. We propose to investigate pavement surface temperatures and pavement heat flux and storage on July 16th and 22nd, when additional mobile surface temperature surveys were conducted. The thermal effects of pavement-watering will be discussed in light of these data series.

Materials and methods

Conductive heat flux and surface temperatures were investigated on Rue du Louvre, near Les Halles in the 1st and 2nd Arrondissements. The street has an aspect ratio approximately equal to one (H/W = 1), is about 20 m wide and has a N-NE – S-SW orientation. It is paved with standard asphalt concrete. Weather station positions are illustrated in Figure 1.

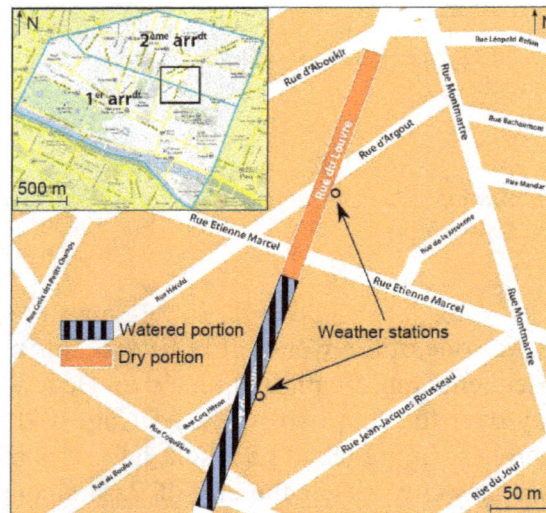

Figure 1. Map of the Rue du Louvre site

These positions were selected as their urban configuration (morphology, orientation, materials, traffic, etc.) was as similar as possible and presented very low vegetation presence. Statistical analyses were conducted using the R software environment, version 3.0.1, retrieved from http://www.r-project.org/ on July 30th, 2013.

Watering method

Watering was started if certain weather conditions, based on Météo-France's three-day forecast, were met. These are summarized in Table 1. Heat-wave warnings are issued if BMI_{min} and BMI_{max} exceed 21 °C and 31 °C, respectively.

Table 1. Weather conditions for pavement-watering

Parameter	Threshold
Mean three-day minimum air temperature (BMI_{min})	$\geq 25\ °C$
Mean three-day maximum air temperature (BMI_{max})	$\geq 16\ °C$
Wind speed	$\leq 10\ km/h$
Sky conditions	Sunny (less than 2 oktas cloud cover)

Cleaning trucks were used to sprinkle approximately 1 l/m² (equivalent to 1 mm) every hour from 6:30 AM to 11:30 AM and every 30 minutes from 2 PM until 6:30 PM on the sidewalk and pavement. A photograph of pavement-watering is shown in Figure 2.

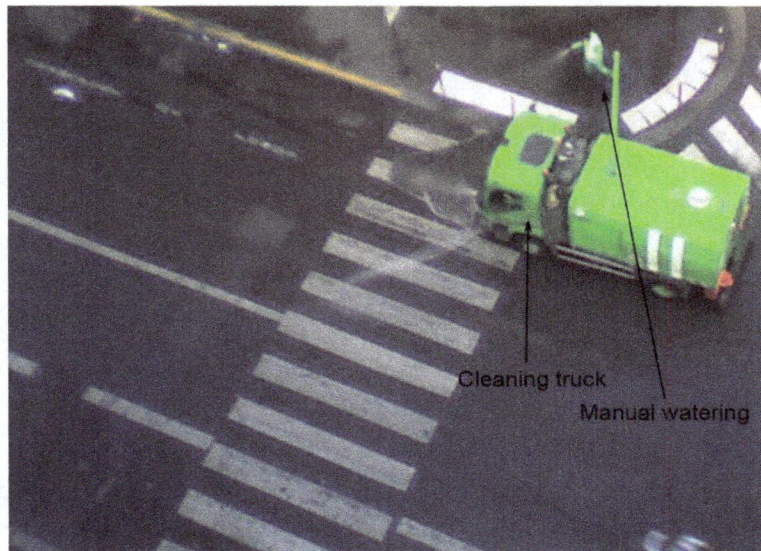

Figure 2. Watering of the pavement and sidewalk on Rue du Louvre

Each watering cycle implies two passes by the cleaning truck. During each pass, the driver waters the two traffic lanes going in his direction, while an operator waters the sidewalk with a high pressure spray. The watering cycle start and end times were reported by the driver. The precision of reporting is estimated to be no better than 5 minutes. The capacity of the cleaning truck is 4 m³, i.e. just 10% more than required for the target area. The cleaning truck therefore refilled between each cycle. The average watering cycle, including watering and refilling, lasted about 20 minutes.

Water used for this experiment was supplied by the city's 1,600 km non-potable water network, principally sourced from the Ourcq Canal. This water network is currently regarded as having high sustainability potential for urban uses that do not require potable water.

It should be noted that applying this watering method to Paris' 2,550 ha of roads and sidewalks would require more than 300,000 m³/day and over 6,000 cleaning trucks. Although this method is clearly unfeasible, it is the simplest way of experimentally watering the road surface while saturating its water-retaining capacity and controlling water consumption.

Pavement sensor

The pavement at each site was equipped with a thermo-flowmeter at a depth of 5 cm, in the centre of the adjoining bus lane. Figure 3 illustrates a top view of sensor installation.

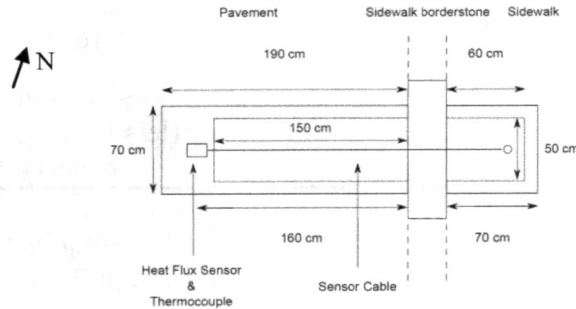

Figure 3. Top view of pavement sensor

In a masked environment such as ours, heat flux sensor positioning determines the exact aspect of the heat flux curve as it defines the beginning and end of pavement insolation, the main contributing factor to pavement heat flux. All else being equal, different positions are therefore not expected to significantly affect our analysis. This should also hold true for streets with similar orientation or aspect ratio. However, differences in paving materials will significantly affect pavement heat flux due to changes in albedo and emissivity. We therefore consider our analysis to be generalizable to all grossly N-S oriented streets with standard asphalt pavement and an aspect ratio of one.

Heat transfer model

Figure 4, based on Kinouchi and Kanda [6], summarizes the heat transfers involved at the pavement surface. Downward solar irradiance is written as S; L_{down} is downward longwave radiation; L_{up} is upward longwave radiation; $S_{reflected}$ is reflected shortwave radiation; H is the upward sensible heat flux; V is the downward pavement heat flux at the surface; ΔQ is the heat storage flux by the first 5 cm layer of pavement; lE is the latent heat flux, with l the latent heat of vaporization and E the evaporation rate. Our flowmeter measures the downward conductive heat flow 5 cm deep, written as G.

Under normal heat-wave conditions, the surface is dry and no latent heat flux is present. The role of pavement-watering is to add a latent heat flux lE as well as an advective heat flux, not shown in Figure 4. The advective flux is created as the sprinkled water heats up from its distribution temperature to the surface temperature of the pavement under wet conditions before evaporating. These two fluxes are expected to cause pavement surface cooling and cascade into micro-climatic effects.

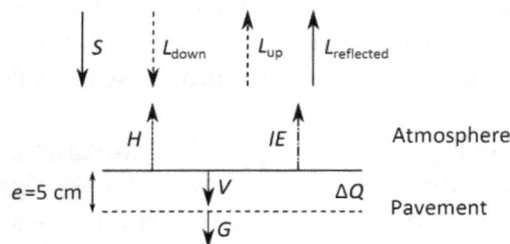

Figure 4. Pavement surface heat budget

Weather station design

The pavement temperature and heat flux sensor was connected to a weather station used for microclimatic measurements. These included air temperature, relative humidity, black globe temperature, wind (velocity and direction), presence of rain and solar irradiance as indicated on Figure 5. Instruments within pedestrian reach were protected behind a cage.

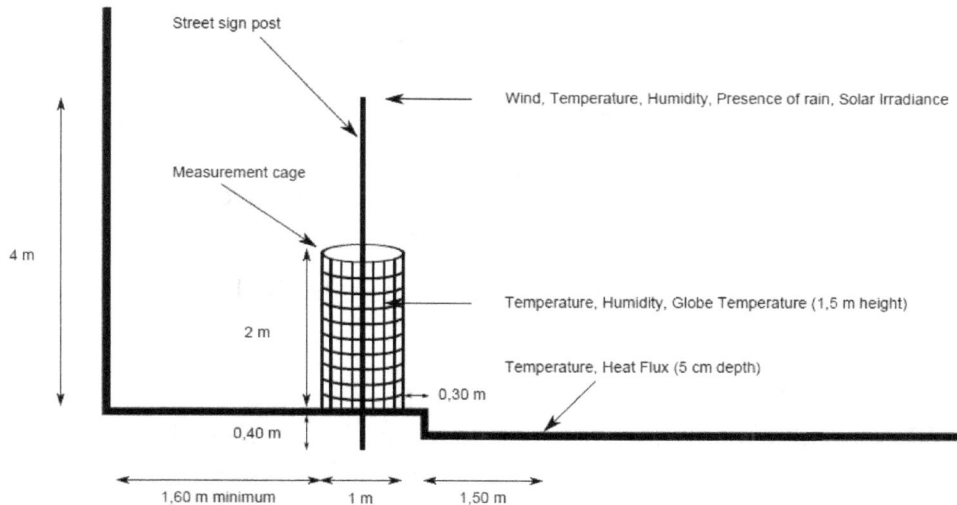

Figure 5. Design of urban weather stations installed Rue du Louvre

Table 2 summarizes the instruments and data used for this analysis. Black globe temperature, wind speed and air temperature were used to calculate mean radiant temperature, as described by ASHRAE [10].

Table 2. Instrument type, measurement height and accuracy

Parameter	Instrument	Height	Accuracy	Parameter
Solar irradiance	Second class pyranometer ISO 9060	4 m	2% daily	Solar irradiance
Pavement heat flux	Taylor-made flowmeter	-5 cm	-	Pavement heat flux
Relative humidity	Capacitive hygrometer	1.5 m	1.5%	Relative humidity
Air temperature	Pt100 1/3 DIN B	1.5 m	0.1 °C	Air temperature
Black globe temperature	Pt100 1/2 DIN A ISO 7726	1.5 m	0.15 °C	Black globe temperature
Wind speed	2D ultrasonic anemometer	4 m	2%	Wind speed

Thermal imaging

A Flir B400 thermal camera was positioned on the roof terrace of 46, Rue du Louvre. The camera recorded false-color IR thermal and visible images simultaneously every hour on non-watered days and every 10 or 15 minutes on watered days. Additional mobile thermal images were taken on site on the evenings of the 16th and 22nd of July with a Fluke TiR32 camera lent by the Parisian Urban Planning Agency (APUR).

These images were used to estimate pavement surface temperature. Apparent (measured) surface temperatures were corrected as presented in Table 3.

Table 3. Parameters used to correct apparent surface temperature

Parameter	Flir B400 (spectral range: 7.5-13 μm)	Fluke TiR32 (spectral range: 7.5-14 μm)
Pavement emissivity	0.97	0.97
Reflected temperature	Mean Radiant Temperature (MRT) as measured by weather station	
Distance to target	20 m	Correction unavailable
Relative humidity	As measured at 1.5 m by weather station	Correction unavailable
Air temperature	As measured at 1.5 m by weather station	Correction unavailable

The emissivity of pavement surfaces was measured in the field using an adhesive of known emissivity.

Results and discussion

Mobile IR thermal photography surveys were conducted with the Fluke TiR32 infrared camera on the 16th and 22nd of July, 2013, from 6:30 PM until 11 PM. Heat flux and storage will also be discussed for these days and compared to those of two control days: July 14th and 20th, respectively.

Infrared thermal photography surveys

Two equivalent zones of pavement, one watered and one control, were studied. Collected data is presented in Figure 6 a, b, as well as surface temperature measurements of an equivalent watered pavement zone made by the Flir camera on the roof-terrace. Fitted regression models for the Fluke data are plotted as well.

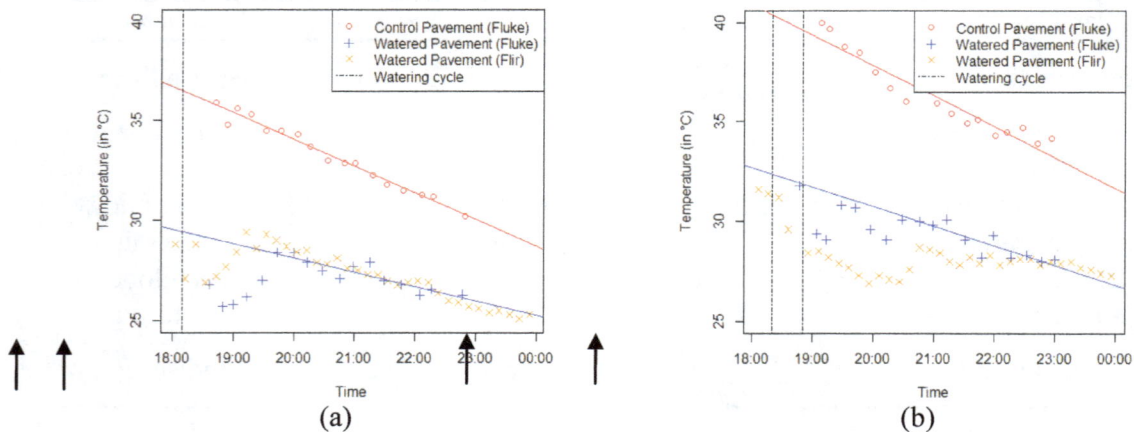

Figure 6. Evolution of corrected pavement surface temperatures: a) 16/07; b) 22/07
Arrows indicate sudden increases or decreases in the Flir data series

The last watering cycle was completed at 6:10 PM on July 16th, while it ended at 6:50 PM on July 22nd. These events are marked with a dash-dotted vertical line. All Fluke thermal images were of shaded areas, while this was not the case for plotted Flir data prior to 6:30 PM.

On both days, Flir measurements are subject to sudden declines and increases between 6 PM and 9 PM. These events are marked in Figure 6 with arrows on the x-axis. The magnitude of these events is about 3 °C on July 16th and 2 °C on July 22nd. These coincide with the high variability seen in the Fluke data before 7:30 PM on July 16th and before 8:30 PM on July 22nd.

These two shifts are respectively caused by pavement shading and surface drying, possibly caused by shortcomings in the applied correction of apparent temperatures. For these reasons, the regressions of the Fluke watered pavement series were only conducted on data collected after pavement drying, i.e. after 7:30 PM on July 16th and after 8:30 PM on July 22nd. The linear model fit was conducted on the whole of the control series.

Table 4 summarizes relevant regression parameters for the control and watered pavements. Given the degrees of freedom, Student tests indicate that the correlations found are at least 99.9% significant on July 16th and July 22nd for both watered and control pavements.

It can therefore be said with high confidence that, over the survey period, the surface temperatures in the control area cool nearly twice as fast as those in the watered area, with a difference of about 7.5 °C at 7:30 PM on July 16th. For July 22nd, which was significantly warmer than July 16th, the control cools nearly 60% faster than the watered pavement, with a difference of about 6 °C at 8:30 PM.

The lower cooling rate of the pavement surface translates lower heat release by the pavement in the hours following pavement drying after watering compared to control conditions. This also indicates lower heat storage by the road materials.

Table 4. Regression parameters

Parameter	July 16th, 2013		July 22nd, 2013	
	Control pavement	Watered pavement	Control pavement	Watered pavement
Slope	-1.33 K/h	-0.71 K/h	-1.54 K/h	-0.98 K/h
R-squared	0.9759	0.8262	0.6052	0.7844
t coefficient	-24.64	-6.896	-12.50	-5.395
Degrees of Freedom	15	10	14	8

Heat flux

Due to difficulties encountered with the control pavement sensor, watered data from July 16th and 22nd will be compared respectively to unwatered data from July 14th and 20th measured at the experimental station. On July 16th, the Météo-France forecast indicated daily minimum and maximum temperatures of 20 °C and 29 °C. On July 22nd, these values were 23 °C and 35 °C, respectively.

Heat flux is highly dependent on solar irradiance. Solar irradiance was 13% lower on July 16th than on July 14th, due to intermittent cloud cover throughout the day. The 4% difference between solar irradiance on July 20th and 22nd is considered negligible. Sunrise occurred at 5:58 AM and 6 AM for July 14th and 16th, while sunset was at 9:52 PM and 9:50 PM, respectively. On July 20th and 22nd, these occurred at 6:04 AM, 6:07 AM, 9:46 PM and 9:44 PM, respectively. Solar irradiance was highest between 1:30 PM and 3 PM.

Figure 7 a, b, c and d illustrate measured heat flux on July 16th, 22nd, 14th and 20th, respectively. Dot-dash vertical lines indicate watering. Control day heat flux ranges between - 50 W/m² and 200 W/m² and follows a skewed-bell-shaped curve. On July 20th, a parked tour bus is responsible for the 30 minute heat flux gap at 2:15 PM. These values are consistent with findings by Asaeda et al. and Kinouchi and Kanda [6, 9].

On watered days however, this shape is very significantly altered. This is especially the case in the afternoon during pavement insolation. In addition, watering cycles coincide well with local heat flux maxima, although this isn't always the case because of intermittent cloud cover.

On July 16th, maximum heat flux is reached when afternoon watering commences, with an approximate value of 110 W/m². On July 22nd; the maximum value of 134 W/m² is also reached a few minutes before afternoon watering begins.

The delay in watering on July 22nd compared to July 16th, due to a delay in the start of pavement-watering, is part of the reason why the observed heat flux maximum is higher, although a passing cloud caused the July 22nd peak to be reached a few minutes before watering. Had there been no interruption of solar irradiance, it is likely that the maximum would have been higher.

For these two days, heat flux maxima are reached during direct sun exposure, just after long interruptions of watering which allowed the surface to dry. The insolation period is that when pavement surface temperatures are expected to be highest. Therefore, it is also the period when advective and evaporative cooling should be highest. These periods, especially that just before the beginning of pavement insolation, therefore appear to be those with the greatest potential to increase the cooling effect.

This suggests that if cities are to carry out large-scale pavement watering, it is crucial to maintain active watering throughout the full duration of the insolation period in order to keep heat flux at a minimum, especially at midday when solar irradiance is highest. In order to avoid heat flux maxima from occuring during pavement watering, a higher watering frequency may be necessary.

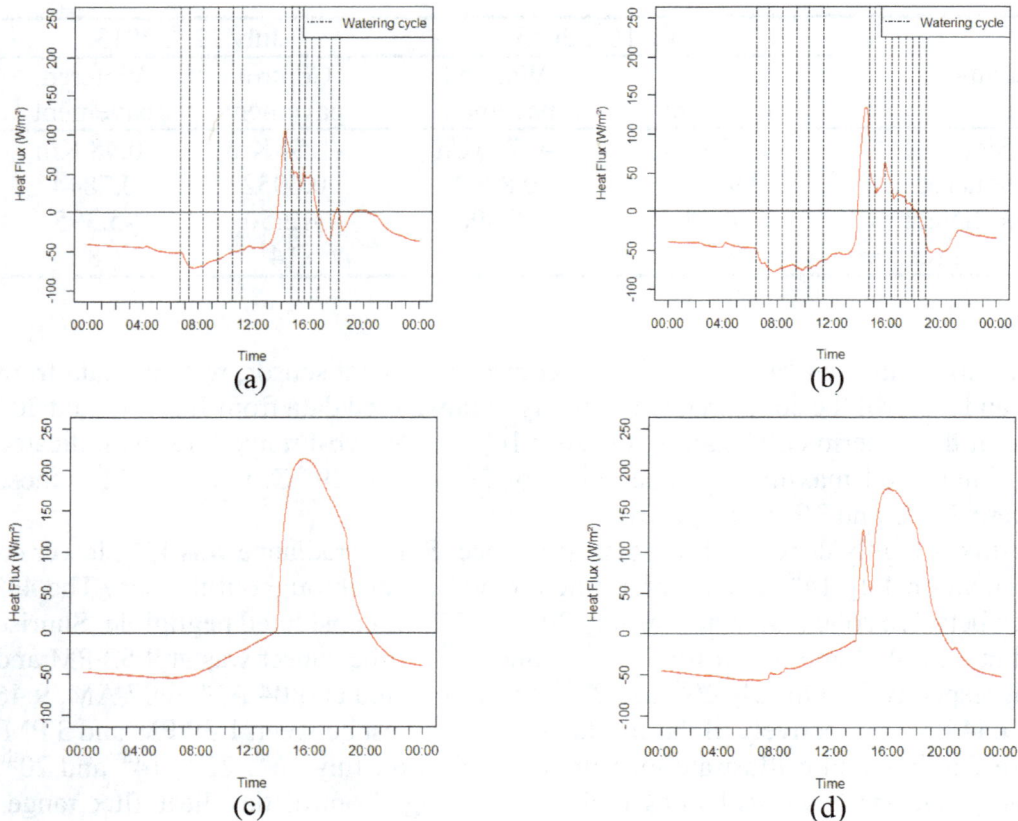

Figure 7. Heat flux measured at 5 cm depth by the 46, Rue du Louvre weather station
a) 16/07; b) 22/07; c) 14/07; d) 20/07

Heat storage

A day-by-day analysis of heat storage will now be conducted. Heat storage was obtained by integration of the measured heat flux by the trapezoidal rule. The analysis will also take a closer look at different periods of the day, defined as follows: morning, the period before pavement insolation, which is divided into two parts: before watering begins, and after watering begins; afternoon is the period of direct sun exposure; evening is the period after direct sun exposure has ended.

Table 5 and Table 6 summarize heat storage on July 14th and 16th, and July 20th and 22nd, respectively. The difference between the sum of the different periods and the indicated total is due to double counting at interval extrema.

For both periods, daily heat storage is significantly reduced: from 0.90 MJ/m² on July 14th to -2.33 MJ/m² on July 16th and from -0.19 MJ/m² on July 20th to -2.81 MJ/m² on July 22nd. The pavement was therefore a net releaser of heat on watered days.

Judging by the details in Table 5 and Table 6, it appears that pavement-watering is most effective at mitigating heat storage during the watering phase, when the latent and advective heat fluxes are present. Furthermore, the effect is maximised during direct insolation of the pavement, when advective and evaporative are expected to be highest.

Table 5. Heat storage (in MJ/m²) on July 14th and 16th, 2013

Date	Morning		Afternoon	Evening	Total
	Before watering	During watering			
July 14th, 2013	-1.25	-0.75	3.15	-0.28	0.90
July 16th, 2013	-1.09	-1.34	0.37	-0.31	-2.33
Difference (Watered-control)	+0.16 +13%	-0.59 -79%	-2.78 -88%	-0.03 -10%	-3.23

Table 6. Heat storage (in MJ/m²) on July 20th and 22nd, 2013

Date	Morning		Afternoon	Evening	Total
	Before watering	During watering			
July 20th, 2013	-1.20	-0.90	2.36	-0.49	-0.19
July 22nd, 2013	-1.01	-1.70	0.52	-0.67	-2.81
Difference (Watered-control)	+0.19 +16%	-0.80 -89%	-1.84 -78%	-0.18 -27%	-2.62

However, because of intermittent cloud cover on July 16th, the observed effect is probably overestimated. Inversely, due to the heat flux gap caused by a parked tour bus on July 20th, the effect observed on July 22nd is probably underestimated.

Overall, if local shading effects are taken into account, it appears that pavement watering is an effective means of reducing heat storage by the pavement below 5 cm. A closer look at the period after pavement drying in the evening would give a better idea of the effects of pavement-watering on nocturnal heat release by road materials.

CONCLUSION

In this paper the effect of pavement watering on surface temperatures and on heat flux and storage 5 cm below the pavement surface was discussed. These aspects were investigated for a street with a N-NE–S-SW orientation and an aspect ratio of one. The road materials were standard, impervious asphalt pavement, a commonly found material in Parisian streets.

Watering was shown to be effective at lowering pavement surface temperatures by several degrees for several hours and significantly reducing the pavement surface cooling rate up to 50% after watering.

Downward pavement heat flux was also found to be significantly reduced by watering. The cooling effect was at its highest during pavement insolation, when advection and evaporation are expected to be greatest.

Finally, heat storage was shown to be significantly lowered, resulting in a negative storage balance for watered days. The reduction in heat storage was highest as watering was taking place, but smaller residual effects were noted after watering had ended as well. As found for pavement heat flux, the greatest effect was achieved during direct insolation of the pavement, although it was not possible to precisely quantify this effect.

From these analyses, it was determined that the greatest potential for improvement of the cooling effect lies in maintaining pavement watering throughout the insolation period, particularly at midday when solar irradiance is highest.

These findings are expected to be generalizable to other streets with similar orientation, aspect ratio and paving materials. However, other configurations and materials should be explored before a city-wide strategy can begin to be elaborated. Further investigation into the micro-climatic effects of pavement watering is also required to ascertain the full benefits of the method. Further research into the optimization of the watering method will help minimize water consumption while maximizing its cooling effects. This will help decision-makers decide whether or not this method is viable at a larger scale.

ACKNOWLEDGEMENTS

The authors would like to thank APUR and Orange for lending their infrared camera and the roof-terrace of their building located at 46, Rue du Louvre at no cost for the purpose of this experiment. They also acknowledge the support of Météo-France and APUR as well as the Green Spaces and Environment, Roads and Traffic and the Waste and Water Divisions of the City of Paris during the preparation phase of this experiment.

Funding for this experiment was provided for by the Water and Sanitation Department of the City of Paris.

NOMENCLATURE

Abbreviations

APUR	Paris Urban Planning Agency
BMI_{max}	Maximum biometeorological index, 3 day mean of daily high temperature
BMI_{min}	Minimum biometeorological index, 3 day mean of daily low temperature
G	Pavement heat flux at 5 cm depth
H	Surface-to-atmosphere sensible heat flux
IR	Infrared Radiation
L_{up}	Upward longwave radiation
L_{down}	Downward longwave radiation
MRT	Mean Radiant Temperature
S	Solar irradiance
$S_{reflected}$	Reflected visible radiation
TEB	Town Energy Balance

UHI   Urban Heat Island
V   Pavement heat flux at surface
WBGT   Wet-Bulb-Globe-Temperature

REFERENCES

1. Robine, J. M., Cheung, S. L. K., Le Roy, S., Van Oyen, H., Griffiths, C., Michel, J. P. and Herrmann, F. R., Death Toll Exceeded 70,000 in Europe during the Summer of 2003., *C. R. Biol.*, Vol. 331, No. 2, pp 171-178, 2008.

2. Météo-France and CSTB, EPICEA - Rapport final, 2012.

3. Masson, V., A Physically-based Scheme for the Urban Energy Budget in Atmospheric Models, *Boundary-Layer Meteorol.*, Vol. 94, No. 3, pp 357-397, 2000.

4. Bouvier, M., Brunner, A. and Aimé, F., Nighttime Watering Streets and Induced Effects on the Surrounding refreshment in Case of Hot Weather, The City of Paris Experimentations, *Tech. Sci. Méthodes*, No. 12, pp 43-55 (In French), 2013.

5. Kinouchi, T., and Kanda, M., An Observation on the Climatic Effect of Watering on Paved Roads, *J. Hydrosci. Hydraul. Eng.*, Vol. 15, No. 1, pp 55-64, 1997.

6. Kinouchi, T. and Kanda, M., Cooling Effect of Watering on Paved Road and Retention in Porous Pavement, in *Second Symposium on Urban Environment*, pp 255-258, 1998.

7. Yamagata, H., Nasu, M., Yoshizawa, M., Miyamoto, A. and Minamiyama, M., Heat Island Mitigation using Water Retentive Pavement sprinkled with reclaimed Wastewater, *Water Sci. Technol. a J. Int. Assoc. Water Pollut. Res.*, Vol. 57, No. 5, pp 763-771, 2008.

8. Takahashi, R. Asakura, A., Koike, K., Himeno, S. and Fujita, S., Using Snow Melting Pipes to verify the Water Sprinkling's Effect over a Wide Area, in *NOVATECH 2010*, pp 10, 2010.

9. Asaeda, T., Ca, V. T. and Wake, A., Heat Storage of Pavement and its Effect on the Lower Atmosphere, *Atmos. Environ.*, Vol. 30, No. 3, pp 413-427, 1996.

10. ASHRAE, *ASHRAE Fundamentals Handbook 2001*, SI Edition, American Society of Heating, Refrigerating, and Air-Conditioning Engineers, 2001.

2

Evaluation of the Trophic Level of Kune and Vain Lagoons in Albania, Using Phytoplankton as a Bioindicator

Anni Koci Kallfa[*], *Fatbardha Babani*, *Ariana Ylli Kraja*
Department of Biotechnology,
Faculty of Natural Sciences, University of Tirana, Albania
e-mail: a_koci@hotmail.com

ABSTRACT

Concentration of chlorophyll is an adequate parameter for assessing the trophic state of lagoon ecosystems. Objectives of this study are: selection of a system of bioindicators to enable a good qualitative evaluation of the trophic state of the lagoons and their dynamics; evaluation of seasonal water quality variability and comparison between lagoons. The trophic state of the lagoons is analysed every month over the year. Water samples are retrieved at four different sites (exact coordinates) each month, sites that are representative of different water circulation systems at each lagoon. The trophic level in the respective lagoons is thus assessed through selection of an adequate system of bioindicators, in order to observe the oscillations of the amount of chlorophyll and therefore to determine the level of eutrophication. Based on the above parameters, the comparison of the trophic state in these two lagoons has shown that they have different trophic states.

KEYWORDS

Kune and Vain lagoons, Lagoon ecosystem, Trophic state, Phytoplankton, Chlorophyll a

INTRODUCTION

The Kune and Vain lagoons of the Adriatic coast of Albania are situated on both sides of Drini river. Due to lack of sufficient maintenance of communication channels, as well as increase of intense agricultural practices along the coastline and around the lagoons, the water quality has worsened and therefore the attention of scientists and managers of natural resources has been focused recently on the assessment of the water quality and the means for improvement. The Kune lagoon communicates with the Adriatic Sea through a 22 m wide strait, while the Vain lagoon communicates with the sea through a 1,300 m long and 20-30 m wide canal, currently closed, and the Zaje-Drin canal, which can be opened to regulate balance between salty and non-salty waters [1].

Data on previous monitoring missions [2] indicates that the trophic levels in these two lagoons significantly differ, since they are separated by Drini river and do not communicate with one another. However, these two lagoons make up one ecosystem, which includes different components, from lagoons to alluvial forests, including agricultural land.

The variation in the trophic level between the two lagoons is thought to be related to the state of communication with the sea, outside sources of pollution [2], and the high amount of evaporation and lack of circulation, especially during summer months.

---

[*] Corresponding author

In the Kune lagoon, station 1 is closest to the communication channel with the sea, while station 4 is the farthest (Figure 1). In Vain lagoon, station 1 is closest to communication with the sea and station 4 is closest to communication with the river that separates the two lagoons. Therefore, based on the approximation to the communication channels, a better trophic situation is expected in sampling stations 1 and 2 in Kune and 1 and 4 in Vain.

KUNE   VAIN

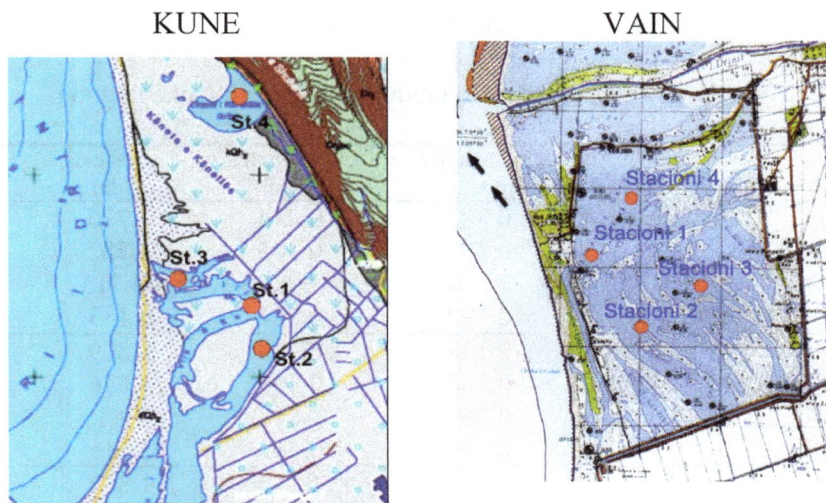

Figure 1. Kune and Vain lagoons and respective sampling stations

METHODOLOGY

The different chlorophyll and non-chlorophyll pigments associated with the photosystems all have different spectra. The content of chlorophyll *a, b, c*, carotenoids, phaeophyta, etc., is determined based on standardized spectrophotometric and fluorimetric methods [3].

The trophic state of the lagoons is analysed every month over the year 2010. Definition of the trophic state as oligotrophic, mesotrophic, eutrophic and hypereutrophic is based on a classification provided in literature. Based on chlorophyll content, the trophic level is grouped into four classes, from lowest to highest: oligotrophic (0—2.6 µg/L), mesotrophic (2.6—20 µg/L), eutrophic (20—56 µg/L) and hypertrophic (56—155+ µg/L) [4]. Based on the data, rehabilitation measures are recommended in order to conserve and improve further the state and the general situation of the ecosystem.

Water samples are retrieved at four different sites (exact coordinates) each month (Figure 1), sites that are representative of different water circulation systems at each lagoon. Pigments are filtered, then extracted with aceton 90%. Determination of pigments is performed based on the trichromatic method, using equations based on the maximums of absorption for each component: chlorophyll *a, b, c* and carotenoids, in a spectrophotometer [5]. Based on the content of photosynthetic pigments, ratios of chlorophyll *a*/chlorophyll *b*, chlorophyll *a*/chlorophyll *c*, chlorophyll *b*/chlorophyll *c* and chlorophyll/carotenoids are calculated, in order to provide indication on the presence of types of microalgae in lagoon waters.

RESULTS AND INTERPRETATION

The trophic state of the lagoons is analysed every month over the year. Water samples are retrieved at four different sites (exact GPS coordinates) each month, sites that are representative of different water circulation systems at each lagoon.

During this study, samples were collected at 4 sampling stations in the Kune lagoon and 4 sampling stations in Vain lagoon over a period of 4 months during year 2010. These stations are chosen such as to represent different water quality and trophic state, mainly based on the distance from communication canals. The exact coordinates are respected during each site monitoring visit. Samples are repetitively retrieved during the period April-October each year [5]. Concentration of chlorophyll *a* and other phytopigments were assessed for each sample (Table 1).

Table 1. Content of photosyntethic pigments in Kune Lagoon

	2010	Chlorophyll *a*, [μg/dm³]	Chlorophyll *b*, [μg/dm³]	Chlorophyll *c*, [μg/dm³]	Carotenoids, [μg/dm³]	Phaeophitins [μg/dm³]
July	St 1	2.073	0.382	0.434	1.776	0.756
	St 2	2.571	0.385	0.305	1.782	0.721
	St 3	81.622	2.173	16.527	46.715	18.246
	St 4	13.628	0.297	1.884	6.133	0.892
August	St 1	2.107	0.148	0.289	1.809	0.531
	St 2	2.155	0.267	0.309	1.646	0.294
	St 3	9.64	0.759	1.817	5.51	0.927
	St 4	3.231	0.514	0.462	2.415	0.275
September	St 1	2.157	0.81	1.112	1.867	0.296
	St 2	2.06	0.213	0.242	1.211	0.097
	St 3	3.079	0.526	0.535	2.438	0.974
	St 4	3.147	0.245	0.314	1.848	0.494
October	St 1	1.794	0.352	0.587	1.433	0.267
	St 2	1.695	0.34	0.515	1.391	0.538
	St 3	2.674	0.34	0.735	2.008	0.168
	St 4	1.532	0.27	0.44	1.173	0.117

Based on previous data [5] in Kune lagoon the average chlorophyll *a* content indicates a deterioration of the trophic conditions from 2002-2007 (Figure 2). This lagoon in general appears to be mesotrophic, but there are peaks (year 2007) when this lagoon has been eutrophic.

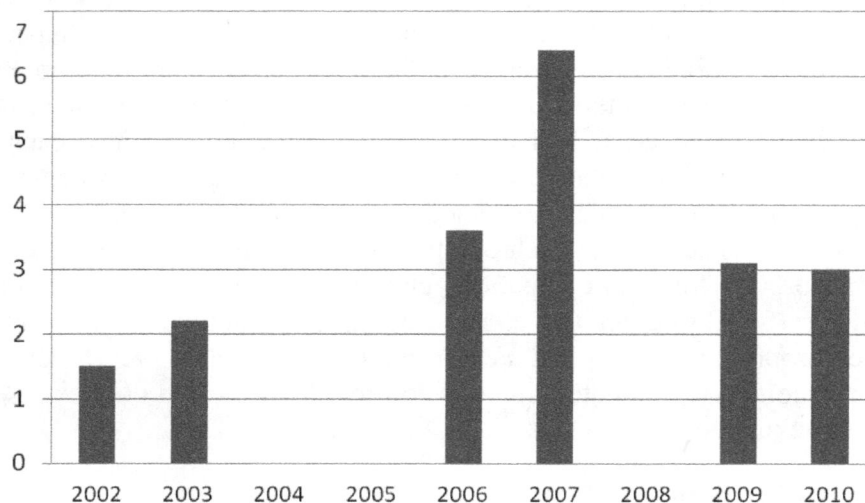

Figure 2. Variation in chlorophyll *a* concentration in mg/m³ over the years at the Kune lagoon [5]

Over the last decade (Figure 2) Kune lagoon initially had a tendency for a trophic state increase in average, with a peak in year 2007, but afterwards the situation has improved. In general, Kune lagoon is in a mesotrophic state.

During 2010, there was noticed a high variation of chlorophyll a content among sampling stations in Kune lagoon (Table 1). Stations 1 and 2 in general appear oligotrophic, while station 3 during the month of August resulted as hypertrophic.

The same applies for Vain lagoon (Figure 3), where trophic conditions have been increasingly deteriorating. In Vain lagoon, based on the data available [5], there may be noticed a tendency for the chlorophyll a content to increase over the years, (Figure 3), resulting in eutrophication.

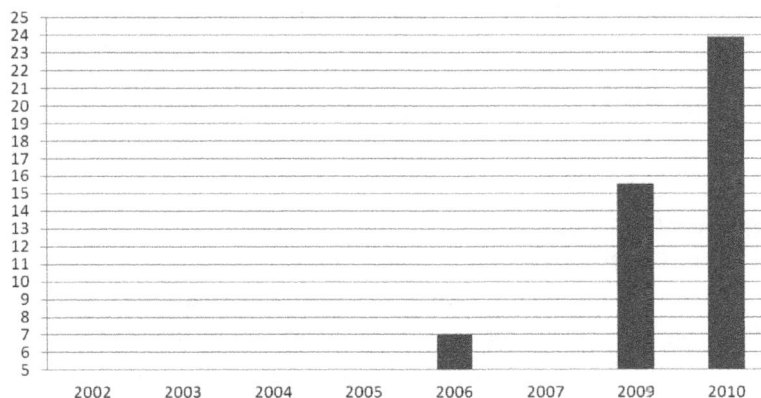

Figure 3. Variation in chlorophyll a concentration in mg/m^3 at the Vain lagoon [5]

Based on previous data, the lagoon of Vain was characterized as oligotrophic during year 2005.

Samples collected during 2010 indicate a high chlorophyll a content and the trophic state has become eutrophic, since the average chlorophyll a content is 24 mg/m^3, while in some sampling stations the trophic state appears also hypertrophic (Table 2).

Table 2. Content of photosyntethic pigments in Vain Lagoon

2010		Chlorophyll a, [μg/dm^3]	Chlorophyll b, [μg/dm^3]	Chlorophyll c, [μg/dm^3]	Carotenoids, [μg/dm^3]	Phaeophitins, [μg/dm^3]
July	St 1	3.396	0.628	0.566	2.603	0.504
	St 2	5.663	0.576	0.66	3.529	1.59
	St 3	6.435	0.83	1.046	3.991	2.029
	St 4	5.893	0.815	0.968	4.945	1.432
August	St 1	12.046	1.006	3.867	6.62	0.722
	St 2	72.249	3.245	11.835	30.509	3.603
	St 3	64.405	5.556	8.187	27.623	15.597
	St 4	60.313	15.106	5.039	27.394	5.885
September	St 1	20.374	0.169	4.361	13.472	0.217
	St 2	28.36	0.122	3.892	15.625	0.48
	St 3	31.495	0.202	5.552	16.644	0.226
	St 4	15.504	0.115	2.179	6.895	0.588
October	St 1	27.099	0.093	4.522	17.495	0.644
	St 2	3.696	0.15	0.434	3.133	0.401
	St 3	5.487	0.094	0.764	3.665	0.748
	St 4	19.692	0.147	3.234	10.884	0.712

In this lagoon, the highest trophic state during 2010 was noticed at station 3 during August (72.3 μg/dm^3).

The distribution of pigments in the algae groups is quite unique. The kinds of chlorophylls are characteristic for each phytoplanktonic family and they can be used as potential taxonomic biomarkers of phytoplankton organisms [6].

Chlorophyll b is characteristic for green algae and chlorophyll c for Diatoms and Dinoflagellates; in Cyanophyceae only chlorophyll a is present [7].

The distribution of chlorophyll a and accessory pigments, chlorophyll b and chlorophyll c, as well as the relative chlorophylls content (ratio of Chl "a" to Chl "b" and to Chl "c") exhibited differences through monitored lagoons.

In Kune Lagoon (Table 1), content of chlorophyll c in the extract indicates the presence of diatoms, while presence of chlorophyll b is evidence for presence of chlorophytes. Since the content of chlorophyll c is higher than the content of chlorophyll b, one can conclude that presence of diatoms is higher than presence of Chlorophytes.

In Vain Lagoon (Table 2), it can be noticed that the content of chlorophyll b and chlorophyll c exhibit the highest values during the month of August. In general, in this lagoon, the content of chlorophyll c is higher than that of chlorophyll b, which again shows higher presence of diatoms.

Values of ratios between chlorophyll a and chlorophyll b and c are high in both lagoons. These ratios keep increasing along the monitoring months.

In the Kune lagoon, during the month of July, it can be noticed that this ratio reaches higher values in Sampling station 3 compared to sampling station 1, where is also noticed a change of status from oligotrophic in station 1 to eutrophic in station 3. Values of this parameter show that the content of chlorophyll a compared to chlorophyll b is high in station 3, which is related to the increase in the trophic level. This indicates that the content of Cyanophyta is very high compared to two other types of microorganisms, which is also connected to the eutrophication nearby this station.

Ratios between chlorophyll a and phaeophytins in both water ecosystems indicate an active state of chlorophylls.

The same conclusions can be drawn based on ratios calculated from sampling data in the two Vain sampling stations.

CONCLUSIONS

Station 2 in Vain Lagoon and station 3 in Kune lagoon are located away from respective communication channels. In these stations are also found the highest values of chlorophyll a content, indicating a high trophic state.

The Kune lagoon is being transformed gradually into a site for throwing wastes. The communicating canal sea–lagoon is completely closed, thus causing a total asphyxia, especially in the summer period that is accompanied with rotting and decomposing of underwater meadows. In this period, because of the high level of evaporation, there is a significant decrease of the level of waters, a high level of eutrophication and an almost total lack of life in it [5].

The lagoon of Vain, being characterized as oligotrophic on average, during the monitoring carried out during 2010, appeared in a eutrophic state at certain stations and months (Table 2). Although it was expected that the trophic state should be better in channels 1 and 4 of Vain lagoon, this was partly true. Trophic state in sampling station 1 of Vain lagoon appeared in a better state compared to other stations, but sampling station 4 that is near the communication channel with the river, at certain months was hypertrophic (Table 2).

This situation requires taking measures for improving circulation of waters, communication with the sea, as well as reducing urban wastes. The factors that may have contributed to the eutrophication of this lagoon are not so obvious. However, it must be emphasized that the communication canals in the Vain lagoon are mostly blocked or in a poor state.

Other factors that may impact the trophic state are population dynamics, increase of agricultural runoff, development of tourism activities, including increase in number of bars and restaurants, lack of waste management in the surrounding area, etc.

Different factors may have contributed to the deterioration of the water quality in these lagoons. In order to draw conclusions on how these factors have directly or indirectly impacted the water quality of this ecosystem, further studies are needed.

RECOMMENDATIONS

The Drini river that divides the two lagoons has an intensive sediment flow, causing changes in the coastal dynamics over the years. This process is still going on, causing the reduction in communication between the Kune lagoon and the sea. Because of circulating current of the side spreading of waves, powerful erosion has developed in the area of coast, which lies to the north of Kune, where besides vegetation of pine trees, the roadway is also being damaged.

Although there is no change in the Vain coastline due to sediments being deposited mainly towards north where Kune lagoon is, the communication canal between this river and the Vain lagoon is not sufficient to ensure good circulation of water. Even the Zaje-Drin canal, the one that connects the Vain lagoon with the Drini river near its delta, is partly blocked and does not ensure sufficient circulation. The Vain lagoon communicates with the sea through the Matkeqe canal, but it is recommended that a new canal be opened north to the existing one, since this lagoon is rapidly reaching high levels of eutrophication.

Based on this study, it results that without immediate measures, no improvement in the water quality of the lagoons will take place naturally.

Threats to the quality of the lagoons include:

- Eutrophication as a result of the reduction of water resources, discharge of sewage water directly to the lagoon, including agricultural run off;
- Reduction in exchange between the sea and the lagoon;
- Reduction and change of water regime of the lagoons as a result of construction of dams along the river deltas;
- Increase in salinity;
- Erosion;
- Uncontrolled population increase ;
- Increased tourism activity;
- Intensification of agricultural activities and as a result increase in nutrient flow.

Proposed measures for improving the situation are several. However, given the specific study, the following measures are recommended:

- Investment in maintenance of canals that connect the lagoon to the sea, especially in Vain lagoon, where there is higher eutrophication;
- Reduction of sewage water discharge, as well as agricultural runoff, limit use of organic or chemical fertilizers nearby the lagoons;
- Evaluation of the impact of different interventions, such as urban growth, increase of pollution or constructions, maintenance of communication canals lagoon-sea;

- Evaluation of the trophic state with the addition of new sampling sites and new indicators.

REFERENCES

1. Miho A., Kashta L., Beqiraj S., *Between the Land and the Sea - Ecoguide to discover the transitional waters of Albania*, Study University of Tirana, 1-462, 2013

2. Anonymous, *Vleresimi i gjendjes trofike dhe cilesise se ujerave ne lagunat e ekosistemit Kune-Vain dhe ne Liqenin e Shkodres*, Universiteti i Tiranes, Fakulteti i Shkencave Natyrore, Departamenti i Bioteknologjise, 2010

3. Lorenzen C. J., Determination of chlorophyll and phaeopigments: spectrophotometric equations, *Limnol. Oceanogr.,* 12, 1967

4. Carlson R. E., Simpson J., Coordinator's Guide to Volunteer Lake Monitoring Methods, *North American Lake Management Society*: 1-96. 1996, (http://dipin.kent.edu/tsi.htm).

5. Jeffrey S. W., Humphrey G. F., New spectrophotometric equation for determining chlorophyll a, b, c1 and c2, *Biochem. Physiol. Pflanz.*, 167, pp 194-204. 1975

6. Schlüter, L., Mohlenberg, F., Havskum, H. & Larsen, S., The use of phytoplankton pigments for identifying and quantifying phytoplankton groups in coastal areas: testing the influence of light and nutrients on pigment/chlorophyll-a ratios, *Marine Ecology Progress   Series   [Mar   Ecol   Prog   Ser]*,   192,   pp   49-63,   2000.

7. Hill, R., Wittingham, C.P., *Photosynthesis*, Edition 2, Publisher Methuen & Company Limited, 1957

3

Raising Energy Efficiency of High-Head Drinking Water Pumping Schemes in Hilly India – Massive Potential, Complex Challenges

Thomas J. Voltz[*1], *Thomas Grischek*[1],
Marcel Spitzner[2], *Jana Kemnitz*[2], *Rudolf Irmscher*[3]
[1]Division of Water Sciences, University of Applied Sciences (HTW) Dresden, Germany
*e–mail: voltz@htw-dresden.de

[2]Faculty of Electrical Engineering, University of Applied Sciences (HTW) Dresden, Germany
[3]Executive Management, Stadtwerke Heidelberg GmbH, Heidelberg, Germany

ABSTRACT

Investigations of energy efficiency of 25 pumps showed wire-to-water efficiencies ranging from 30% to 60%, with an average of 47%. Raising the efficiency of just 7 pumps to the realistic target of 60% would require an initial investment of 126 k€ and represent a net present value (profit) of 446 k€ over a 10-year pump lifetime, saving 8.6 kt of CO_2 emissions. The primary measures for raising efficiency are in order of priority: 1) improving pre-filtration of raw water to prevent rapid mechanical wear due to suspended particles during monsoon, 2) providing training, improved working conditions, and better tools and spare parts among pump operators and 3) replacing aging, oversized pumps with properly sized pumps operating close to peak efficiency. As of January 2014 the results have been confirmed by a Bureau of Energy Efficiency-certified energy auditor and the extent and funding of efficiency measures implementation is in planning.

KEYWORDS

Energy efficiency, Pumps, Drinking water, India, Capacity building, Emission reduction

INTRODUCTION

Pumps make up 20% of the global electrical energy demand, and 60 to 95% of that of drinking water supply systems [1-3]. This use of energy represents 70 to 90% of the life cycle cost of a pump, depending on the required head and flow as well as cost of maintenance and personnel [4-6]. By comparison, the up-front purchase cost of a new pump tends to lie between 3 and 10% of its life cycle costs [7]. This underlines the importance of stressing high energy efficiency in the selection of a pump, which ought to be the second priority after meeting the drinking water demand. Pumping systems are known to have a significant potential for energy efficiency improvements [8-12]. This is primarily because it is rather easy to select and operate pumps at low efficiency through conservative over dimensioning, changing demand requirements, inefficient flow control (bypass or throttle rather than variable speed operation) [13, 14], inefficient pump scheduling [15] and negligent or poor maintenance [16]. The energy and cost savings that stand to be gained through modification or replacement pumping systems are estimated to be as high as 30% [8].

---

* Corresponding author

Although according to the literature cited, this topic is well known and well researched in Europe and North America, in India this appears not to be the case, despite what is likely a large potential. Attention has been given to irrigation pumping systems and their energy efficiency [17, 18], but scientific studies specifically on drinking water pumping systems are scarce. Past experiences in India [19] and elsewhere have also shown that many projects fail to succeed due to problems with an agreeable funding solution and effective execution by a team composed of private consultants and public water supply company engineers. This paper will analyse the example of the drinking water supply pump systems of the hill cities of Srinagar/Pauri and Mussoorie in Uttarakhand, India. The goal is to illustrate 1) the current state of energy efficiency, 2) the reasons for low efficiency and the corresponding approaches to improving it and 3) the potential for energy and cost savings and thoughts on how to best achieve them.

DRINKING WATER SUPPLY IN UTTARAKHAND, INDIA

In the hill regions of the small North-Indian state of Uttarakhand (Lower and Middle Himalaya), drinking water supply is almost exclusively based on surface water. Most schemes for small villages are gravity-fed, but approximately 30 large towns (>5,000 inhabitants) require additional input of energy via electric motor-driven pumps to lift the water to the point of consumption. At the two study areas Srinagar/Pauri and Mussoorie horizontal multistage centrifugal pumps are used to lift water as much as 1,700 m from rivers and streams to storage tanks, requiring multiple boosting stations and working heads up to 600 m for moderate flows up to 120 m^3/h. As individual pump-motor sets have real power demands as high as 200 kW, pumping is very energy intensive and comprises >90% of the electricity demand. The partner water supply organization Uttarakhand Jal Sansthan (UJS) pays app. 1 million €/year in electrical energy bills for pumping at these two areas alone.

Figures 1 and 2 show the central raw water extraction station for the hill city Srinagar, through which 70,000 people are supplied with approximately 10,000 m^3/day of drinking water. The existing pump sets are exposed to the damaging effects of nature, chief among them the annually recurring monsoon with high-turbidity floods of varying intensities. The potential for improvement is evident.

Figure 1. Single-stage raw water pumps installed on an open platform in Srinagar

Figure 2. Suction pipes with non-return valves in the River Alaknanda

At the second study area in the hill city of Mussoorie, staff and students of the University of Applied Sciences Dresden, supported by the contracted Indian certified energy auditor, inspected 45 multistage pumps, of which 25 were in operating condition. Measurements of flow, pressure rise across the pump (difference between discharge and suction pressures) and electrical power input yielded a mean "wire-to-water" energy efficiency of 47%, with values ranging from 30 to 60%. Of the 22 pumps, 10 were over dimensioned by a margin greater than 15%, meaning the rated head (as on the nameplate) was 15% greater than the actual working head measured during operation using an analog pressure gauge. The appearance of the pump and pump house was often an unreliable indicator of efficiency (Figure 3), such that examples of relatively efficient pumps were also found in unexpected places (Figure 4).

Figure 3. Pump set 1 at Jinsy II Pump Station, wire-to-water efficiency of 47%, Mussoorie

Figure 4. Pump set 6 at Murray Pump Station, with an efficiency of nearly 60%

MEASURES FOR AND CHALLENGES AGAINST RAISING EFFICIENCY

Experience gained during measurements has revealed three primary measures necessary for raising the efficiency and the salient challenges that must be overcome to implement these measures.

Pre-treatment

In many cases there is inefficient or non-existent raw water pre-treatment through physical filtration to remove suspended particles (silt and sand). This problem is especially severe during monsoon. The internal components of the pumps, which due to high head requirements operate predominantly at rotational speeds of 3,000 min^{-1}, are heavily eroded and rapidly lose efficiency. For example, the raw water pumps in Figure 1 are normally replaced or completely overhauled every year after monsoon. After replacement, they have a total efficiency of 60%, but after one year drop to an average of about 30%. This fact is well known among UJS engineers and explains the ubiquitous over dimensioning, since the pumps would otherwise be unable to meet their operating requirements after just a few years, as has been the case several times in the past. Many "clear water" pumps at sites in Mussoorie are also likely affected, which due to their high energy consumption compared to the raw water pumps is even more costly. This problem can be addressed by superior intake structures (Srinagar) or improved filtration and sedimentation (Mussoorie) before the pump inlets as well as erosion-resistant impeller materials and surface coatings.

The main challenge against improving this pre-filtration is designing for the extreme flow and high shear stress conditions that develop during monsoon, which require structures with well-anchored foundations and a robust superstructure. This requires a complex engineering solution and large initial investments, which according to UJS engineers are very difficult to obtain. This is due to a culture of short-term planning and brought about to a large extent by the difficult climatic conditions, which cause extensive damage to water supply schemes each year and keep the focus on the many recurring repairs, rather than allowing for carefully planned schemes with longer lifetimes and higher initial investment requirements.

Training and equipment of pump operators

Most pump operators possess a low level of training and rely solely on practical experiences gained on the job. Their operation and maintenance methods often achieve baseline effectiveness, but far from the technical standards common for example in Europe. The working conditions with respect to safety (risk of falls, noise level, room lighting) are generally in need of improvement. Special tools are rarely available, such that complex tasks such as exact alignment of the pump and motor cannot be correctly executed. High-quality spare parts such as pump bearings and impellers are more expensive and difficult to obtain, such that lower-quality and less-expensive local options offer the only feasible solutions to urgent repair needs. These problems can best be addressed through intensive and regular training programs for the pump operating staff with focus on pump installation and maintenance. This must be accompanied by an improvement in the pump station working conditions and availability of appropriate tools and selected spare replacement parts. In this way, higher energy efficiency would be a positive side-effect.

Challenges to this include the general opinion that investments in the training of low-level workers are not very beneficial, and difficulty in overcoming the institutional inertia to change the status quo. In some cases, such as in Srinagar, the pump staff are outsourced from other companies and employed on short-term contracts, a fact which speaks against investing in their education. In India it is common for lower-level employees to display limited loyalty to their companies and to seek and pursue the best economic opportunity available.

Systematic pump replacement

Once the first measures have been implemented, the pumps should be systematically replaced, beginning with the oldest, least efficient pumps with the highest electric power consumption and operating hours. The selection of new pumps must be done properly to avoid unnecessary over dimensioning. Similarly, the operating regime must be taken into account in the pump controls, for example such that in the case of widely varying pump duty requirements the efficiency is kept at a maximum. This can be accomplished for example by a variable frequency drive in place of a typical throttle valve, or an effective parallel operation of one high-flow and one low-flow pump in place of two identical pumps.

Other technical measures identified by the certified energy auditor [20] also show promise, such as compensating for low power factors using capacitor banks, adjusting the contracted maximum electrical load (in kW), and drawing power at higher voltage. These proposed measures will be included in the planning for the implementation of energy efficiency improvement measures in 2014.

POTENTIAL FOR COST SAVINGS AND CO_2 EMISSION REDUCTION

Pump replacement vs. continued operation of existing pumps

To illustrate the savings potential for all 25 pumps measured in Mussoorie, a conservative cost-benefit analysis was conducted using a dynamic cost comparison. In the improved case, it was assumed that all pumps were replaced with new pumps from the German manufacturer KSB (also widely available in India), which were chosen using their proprietary internet-based tool "KSB EasySelect" according to the operating requirements measured with open throttle valve. The pump lifetime (running time without complete replacement or major repair) was set to a maximum of 20 years and a specific price per capacity (€/[m×l/min]) for purchase, installation and of the new pump, which varied as a falling power function. The means that the specific price decreases according to the following function:

$$Specif ic\ Costs\ in\ Rs.\ (INR)\ =\ 17.07\ \times\ (Head\ \times\ Flow)^{-0.4508} \qquad (1)$$

This was based on recent price quotes for 7 pumps received from UJS. As an example, a pump with a required flow of 1,200 l/min and a head of 315 m would have a specific investment cost of 0.0522 €/(m×l/min) and incur a total investment cost of 19,750 €. It was assumed that the efficiency of the new pumps would linearly degrade throughout their lifetime, such that in the final year of operation they reach the current efficiency of the pump to be replaced. The initial electricity price was 5.4 ct €/kWh with a rapid price increase of 13.3%/year (based on pump station electricity bills from 2010 to 2013). A nominal interest rate of 7.75% was used (based on World Bank data). In the base case, it was assumed that the all pumps would continue to be operated with an unchanging efficiency. These two cases were compared based on their net present values.

Figure 5 shows the progression of the net present values for each pump with respect to pump lifetime, calculated by subtracting the net present value of the base case from that of the improved case. The exponentially rising curves show the influence of the very high electricity price increase and make it clear that a long lifetime (for example through improved maintenance) can yield especially large cost savings. The replacement of 7 pumps (black solid lines) leads to particularly good economic viability owing to their high operating hours and electrical power consumption, and results in payback periods (intersection of the net present value lines with the x-axis) from 4.5 years to 1 year. Only 4 of 25 the replacements do not pay for themselves within the simulated maximum 20-year pump lifetime.

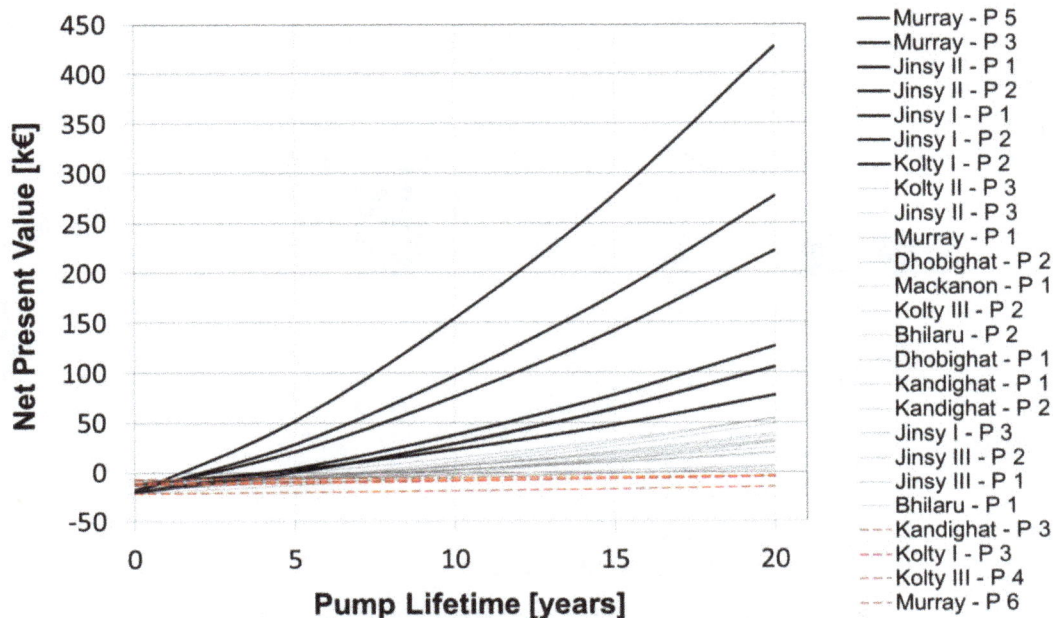

Figure 5. Economic viability of the replacement of the existing 22 pumps in Mussoorie with new pumps, as shown by the difference in net present value between pump replacement (improved case) and continued operation of existing pumps (base case). k€ = thousand Euro

Emission reduction

Improvements in pumping energy efficiency can yield not only large cost savings, but also substantial CO_2 emissions reduction from imported electricity from outside the state of Uttarakhand, of which 67% is generated using fossil fuels (Figure 6). Based on figures from the Central Electricity Authority [24] for tons of CO_2 output per MWh of coal (1.08 t/MWh) and gas (0.46 t/MWh) burned in North Indian power plants, and the mixture of imported fossil-based electricity into Uttarakhand (55% coal, 12% gas), the conversion factor is 0.968 t/MWh. This assumes that only fossil-based electricity is replaced by energy efficiency improvements, as per a federal policy that favours the generation from other sources such as hydropower.

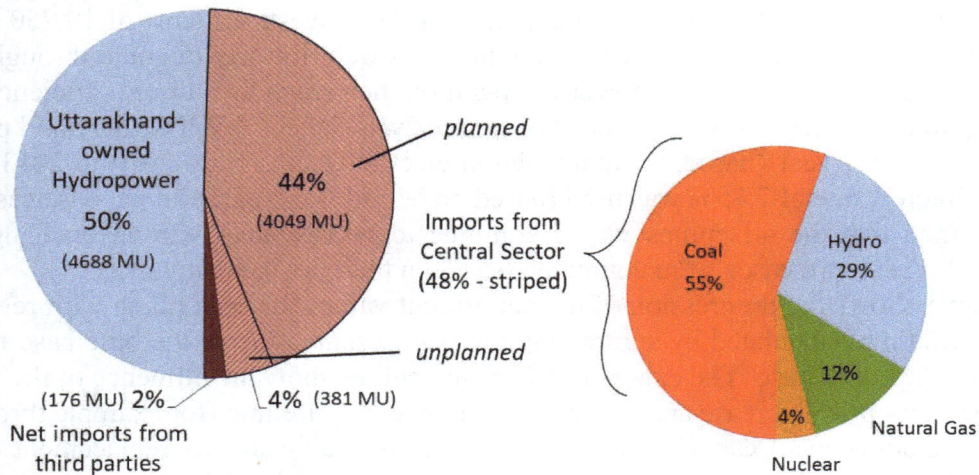

Figure 6. Electrical energy quantities generated in Uttarakhand, 2010-11, broken down according to fuel type and source of supply, whether in-state or from neighboring states (central sector). MU (Million Units) = GWh [21-23]

Total savings

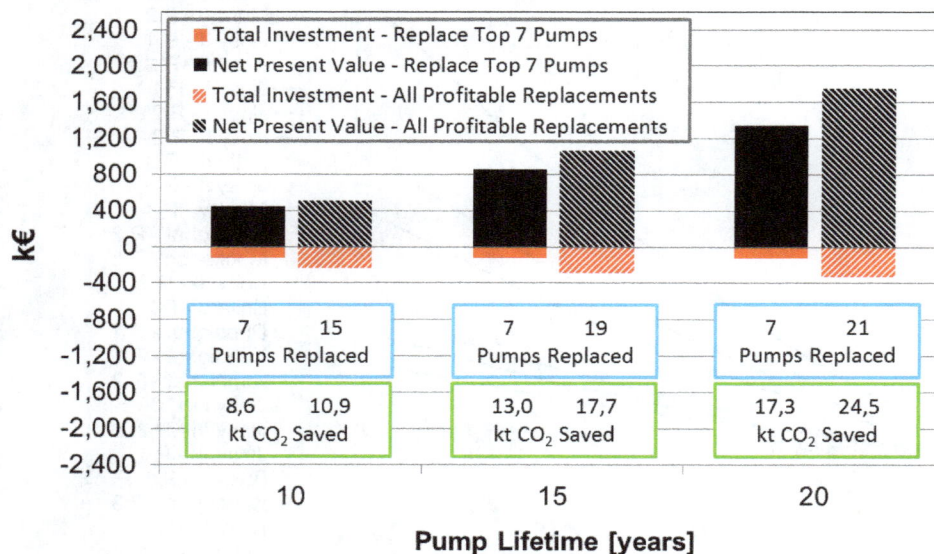

Figure 7. Net present values and total investment costs for pump replacement under various assumptions of the lifetime for 7 pumps and all pumps, respectively, for Mussoorie. k€ = thousand Euro, kt = thousand tons

The cost savings and CO_2 emission reduction possible through the replacement of just 7 pumps are shown in Figure 7 for three different pump lifetimes. The investment of funds should always be made for those pumps, which yield the greatest returns. For example, the investment of 126 k€ (k€ = thousand Euro) for the replacement of 7 pumps leads to a savings (profit) of 446 k€ after a 10-year lifetime versus continued operation (base case), a rate of return of 353%. This results in a CO_2 emission reduction of 8.6 kt (kt = thousand tons). By contrast, a nearly twofold investment of 241 k€ for the replacement of 17 pumps after a 10-year lifetime yields a savings of 507 k€ and a rate of return of (only) 210%.

CONCLUSION AND THOUGHTS ON FINANCING THE IMPLEMENTATION

The authors recommend that UJS undertake a pilot energy efficiency project to replace some or all of the 7 pumps in Mussoorie showing the greatest potential savings. The number to be replaced depends on the amount of funding that can be acquired externally or freed up internally by UJS to cover initial investment costs. Based on discussions and experiences thus far, external funding will be necessary, and is in the process of being applied for with the aid of a state-level funding agency (Uttarakhand Renewable Energy Development Agency, UREDA), once these first measurements of energy efficiency are confirmed by an accredited energy auditor. As UJS has limited additional engineering capacity to oversee the technical implementation of this project, which is of secondary importance compared to effective water supply, continued support from a third-party will be required.

In the best case this would be done by an energy service company (ESCO) that provides investment funding and engineering services and later receives payment on the basis of performance contracting. This means that the ESCO receives its payment as a percentage of the cost savings achieved each year for a contractually agreed-upon period following the implementation of the efficiency measures. While such companies are active in North India, they are more hesitant to enter into such contracts with municipal organizations, which face various challenges that could affect regular payments. For this, a second third-party company or bank may be required to broker the payments via a special account.

UREDA itself could act as a kind of non-profit ESCO by providing a small amount of funding as an interest-free loan, which could suffice for the replacement of 2 or 3 pumps. Alternatively, although less desirable in the eyes of the authors, is the possibility of receiving a low-interest loan from the Indian Renewable Energy Development Agency (IREDA), which carries a grace period up to 3 years before repayment begins. This should allow for sufficient time to reap the benefits of the improved efficiency and enable timely repayment without additional financial burden on the UJS budget. However, in both cases an external engineering firm in India will need to be separately arranged to supervise the implementation. These options will be explored in the coming months such that implementation of at least pilot-scale measures can be achieved by the end of 2014.

ACKNOWLEDGEMENTS

The encouragement and organisational support of Mr. A.K. Tyagi of UREDA, the technical support of the staff of UJS, especially Mr. P.C. Kimothi, Mr. L.K. Adlakha, Mr. S.K. Sharma and Mr. Rohela, the electro technical consultation of Mr. S. Dhamija of Simplex Control Systems Co., and the advice and field work of Mr. Rakesh Yecho of Technical & Management Consultancy Center (subcontractor of PCRA) and his team are gratefully acknowledged. Funding of the project was provided by the Stadtwerke

Heidelberg GmbH in agreement with the certification agency GSL (Green Electricity Label). Additional support was received through the DAAD programme "A new passage to India 2013-2014" (grant no. 56040107).

REFERENCES

1. Europump and Hydraulic Institute, Pump Life Cycle Costs: A Guide to LCC Analysis for Pumping Systems, 1st Edition, Ed. Lars Frenning. Parsippany, New Jersey, 2001.
2. Schweizericher Verein des Gas - und Wasserfaches (SVGW) und EnergieSchweiz, Energy in the water supply - Guide to energy cost and operational optimization, Zürich, 2004. (in German)
3. de Almeida, A., Fonseca, P., Falkner, H., Bertoldi, P., Market transformation of energy-efficient motor technologies in the EU, *Energy Policy*, 31(6), pp. 563-575, 2003.

4. Electric Power Research Institute (E PRI), *Program on Technology Innovation: Electric Efficiency through Water Supply – A Roadmap*, Palo Alto, California, USA, 2009.
5. Deutsche Energie-Agentur GmbH (DENA), Pumps and pump systems gudie for industry and commerce, *Pumps and pump systems guides for industry and commerce,* Initative EnergieEffizienz, Berlin, 2010. (in German)
6. Global Water Research Coalition (GWRC), *Energy Efficiency in the Water Industry: A Compendium of Best Practices and Case Studies – Global Report*, International Water Association (IWA) Publishing, London and New York, 2012.
7. Dickenson, T. C., *Pumping Manual*, Oxford, UK: Elsevier Advanced Technology, 1995.
8. Kaya, D., Yagmur, E. A., Yigit, K. S., Kilic, F. C., Eren, A. S. and Celik, C., Energy efficiency in pumps, *Energy Convers Manag*, 49(6), pp. 1662–73, 2008.

9. Ferreira, F. J. T. E., Fong, C., and de Almeida, T., Ecoanalysis of variable-speed drives for flow regulation in pumping systems. *IEEE Transactions on Industrial Electronics*, 58(6), pp. 2117-2125, 2011.
10. Pemberton, M., and Bachmann, J., Pump systems performance impacts multiple bottom lines, *Engineering & Mining Journal*, 211(3), pp. 56-59, 2010.
11. Zhang, H., Xia, X., and Zhang, J., Optimal sizing and operation of pumping systems to achieve energy efficiency and load shifting, *Electric Power Systems Research*, 86, pp. 41-50, 2012.
12. Vilanova, M. R. N. and Balestieri, J. A. P., Energy and hydraulic efficiency in conventional water supply systems, *Renew Sustain Energy Rev,* 30, pp. 701-714, 2014.

13. Europump and Hydraulic Institute, *Variable Speed Pumping*: *A Guide to Successful Applications*, 1st Edition, Cornwall, Great Britain, 2004.
14. Viholainen, J., Tamminen, J., Ahonen, T., Ahola, J., Vakkilainen, E., Soukka, R., Energy-efficient control strategy for variable speed-driven parallel pumping systems, *Energy Efficiency*, 6(3), pp. 495-509, 2013.
15. Bagirov, A. M., Barton, A. F., Mala-Jetmarova, H., Al Nuaimat, A., Ahmed S. T., Sultanova, N., Yearwood, J., An algorithm for minimization of pumping costs in water distribution systems using a novel approach to pump scheduling, *Mathematical and Computer Modelling*, 57, pp. 873-886, 2013.

16. Feldman, M., Aspects of energy efficiency in water supply systems, *Proceedings of the 5th IWA Water Loss Reduction Specialist Conference*, South Africa, 2009.
17. Purohit P. and Kandpal, C., Renewable energy technologies for irrigation water pumping in India: projected levels of dissemination, energy delivery and investment requirements

using available diffusion models, *Renew Sustain Energy Rev*, 9(6), pp. 592-607, 2005.

18. Shah, T., Scott, C., Kishore, A., Sharma, A., Energy-Irrigation Nexus in South Asia: Improving Groundwater Conservation and Power Sector Viability, Second (Revised) Edition, *International Water Management Institute,* Colombo, Sri Lanka, 2004.

19. Bureau of Energy Efficiency (BEE), Alliance to Save Energy & International Finance Corporation (IFC), *INDIA: Manual for the Development of Municipal Energy Efficiency Projects*, Copyright 2007 by IFC, New Delhi, 2008.

20. Petroleum Conservation Research Association (PCRA), *Energy Audits of Drinking Water Pumping Stations in Uttarakhand, June/November 2013*, on Behalf of the Uttarakhand Renewable Energy Development Agency (UREDA), January 2014, Delhi.

21. Central Electricity Authority (CEA), *Operation Performance of Generating Stations in the Country during the Year 2010-11*, Operating Performance Monitoring Division, April 2011, Delhi, 2011.

22. Northern Regional Load Despatch Center (NRLDC), *Daily Power System Energy Reports, April 2010 to March 2011*, Power System Operation Corporation Limited, Delhi, 2011a.

23. Northern Regional Load Despatch Center (NRLDC), *Annual Report 2010-11*, Power System Operation Corporation Limited, Delhi, 2011b.
    Central Electricity Authority (CEA), CO_2 Baseline Database for the Indian Power Sector, User Guide. November 2006, Delhi, http://www.cea.nic.in /reports/planning/-cdm_co2/final_user_guide.pdf, [Accessed: 04-Jun-2012]

4

Electrification using Decentralized Micro Hydropower Plants in North-eastern Afghanistan

Ramchandra Bhandari[*1], *Anita Richter*[2], *Andre Möller*[3], *Rolf.-P. Oswianoski*[4]

[1]Institute for Technology and Resources Management in the Tropics and Subtropics,
Cologne University of Applied Sciences,
Betzdorfer Strasse 2, 50679 Cologne, Germany
e-mail: bhandariramchandra@yahoo.com
[2]Deutsche Gesellschaft für Internationale Zusammenarbeit (GIZ) GmbH,
Potsdamer Platz 10, 10785 Berlin, Germany
[3]Alteburger Strasse 109, 50678 Cologne, Germany
[4]Virtu Consult UG, Handjerystraße 94, 12159 Berlin, Germany

ABSTRACT

Electricity supply database from the Afghan national authority for electricity supply shows that about only 20% of the population in Afghanistan had access to grid electricity by 2010. The national utility has a total capacity of about 842 MW, out of which about 696 MW was operational. Additionally, many decentralized units (Micro Hydropower (MHP) plants, diesel generators and solar home systems) supply electricity to about 7% of the population. The donors supported National Solidarity Programme (NSP) has promoted hundreds of rural electrification projects. MHP plants are the major renewable energy based projects among them. In order to identify the operational status of installed MHP plants in four North-eastern provinces (i.e. Badakhshan, Baghlan, Balkh and Takhar) and to assess their socio-economic impacts, an extensive field monitoring had been carried out. The major parameters studied were spatial distribution of MHP plants, investment costs, operational models, end user electricity tariffs, productive use of electricity, community satisfaction, etc. Altogether, 421 MHP installations (about 11 MW installed capacity) were visited. The outcomes obtained from those surveys are presented in detail in this paper.

KEYWORDS

Electricity supply, Field survey, Hydropower, Renewable energy, Rural electrification, Productive use.

BACKGROUND

Energy is one of the basic inputs for all economic activity. Per capita energy consumption is one of the major determinants as well as an indicator of economic development. Per capita energy consumption and economic growth reinforce each other [1-3]. Access to modern energy, particularly to electricity, cuts across all sectors in rural development. Although access to electricity is not an end in itself, it is an essential tool to facilitate social and economic activity [4].

Grid electricity is available mainly in the big cities in Afghanistan. Most of this supply comes from imports, while the rest is generated within the country from hydro and thermal (diesel) resources. By 2010, the country had about 842 MW installed capacity.

---

[*] Corresponding author

Out of this, about 696 MW was reported to be operational [5]. In 2011, about 3,086 GWh electricity was supplied by the national utility "Da Afghanestan Breshna Sherkat (DABS)" in the country. The share of grid electricity supply by source types in 2010 - 2011 is given in Table 1.

Table 1. Electricity generation in Afghanistan by source types in 2010-11 [6, 7]

Source types	2010 [GWh]	2010 share [%]	2011 [GWh]	2011 share [%]
Hydro	909	35.2	802	26
Thermal (diesel)	101	3.9	40	1.3
Import	1,574	60.9	2,244	72.7
Total	2,586	100	3,086	100

As shown in Table 1, the country depends largely on imported electricity from its neighbours (mainly from Turkmenistan, Tajikistan, Uzbekistan and Iran). Hydropower is the major source for inland electricity generation. Thermal generation is mainly based on imported diesel. The share of other renewable energies (e.g. solar, wind, biomass, etc.) for electricity supply is negligible.

According to the Afghan Central Statistics Organization (CSO) [8] estimate, the total population of the country (in its 34 provinces) was around 24 million in 2009/10. About 76% of this population lives in rural areas [9], mostly without access to modern form of energy. Table 2 shows the DABS customer numbers under different user categories from 2006 to 2010 throughout the country.

Table 2. Numbers of DABS customers (connections) in Afghanistan [10]

Year	2006	2007	2008	2009	2010
Residential	391,683	502,856	574,379	671,266	682,454
Commercial	30,888	39,811	47,139	56,803	58,381
Government	3,187	4,208	4,950	4,153	4,191
Holy places	4,449	2,699	3,941	5,788	5,921
Industrial	2,399	1,688	2,620	1,912	1,860

Not surprisingly, the residential customers overshadow the other customer groups. The specific information on the amount of electricity consumed by each group was not available. An unofficial estimate from the DABS implied that about 80% of the sold electricity is consumed in residential sector, about 10% in industrial sector and the rest 10% in other sectors. Assuming an average of 7 people per customer (household) in residential sector [11], it can be said that about 20% of the country's population was supplied with the grid electricity in 2010. Additionally, an estimated 12% of the rural population in the country would have access to electricity, provided all the installations under National Solidarity Programme (NSP) would have been operational (Figure 1). The rest of the population depends heavily on kerosene lamps and fuel wood for lighting and on diesel for operation of agricultural machinery such as flour grinding mills, etc. Direct burning of biomass (agricultural residues and fuel wood) is the major source for cooking and space heating energy needs. The households and commercial entities connected to the grid do not have a reliable supply. Scheduled load shedding and unscheduled blackout is common. No official data could be obtained from DABS regarding the average power-cut hours in a year, however, the residents reported it to be about 4-6 hours a day.

Private sector development in the area of rural electrification through renewable energies is still weak in the country. Security risks for investment and low electricity demand as well as purchasing power of the rural population are two of many reasons. At present, the rural electrification through renewable energy strongly relies on government and foreign donor supports. Due to factors such as rough terrain, extreme remoteness, high investment costs, etc. among many others, the national grid extension throughout the country is very unlikely in the foreseeable future. Therefore, off grid alternatives offer a more realistic option when it comes to increasing the coverage of electricity access in developing regions [12]. Further installations and proper monitoring of the isolated power generation systems, e.g. MHP plants and solar home systems (SHSs), could contribute to the electricity supply to rural folks.

There are significant numbers of smaller projects implemented for rural electrification in Afghanistan (mainly under the NSP). The sustainability of any rural electrification project requires active local participation in development and implementation of such projects [13]. The NSP serves this purpose. It is a nationwide rural development program first established in 2003. The program states its goal as to develop the ability of Afghan communities to identify, plan, manage and monitor their own development projects [14]. The program's main objective is the establishment of community development councils (CDCs) to function as institutions for local governance leading the social and economic developments in their communities. The CDCs lead national efforts for local governance, rural development and poverty alleviation in their communities [15].

As of December 2010, there were as many as 51,870 projects implemented under the program with a total cost of USD 750 million [16]. They include both infrastructure projects such as water and sanitation, power, irrigation, transport, and education as well as social projects like vocational training and literacy courses. Energy sector is an important component of the program and there were as many as 6,680 energy projects with a total cost of USD 148 million implemented under this program [16]. Renewable energy based rural electrification projects implemented under NSP include small wind turbines, MHP plants, SHSs, etc. [15]. In the past diesel generators were also promoted intensively; however, NSP does not support diesel plants anymore. The capacity installed

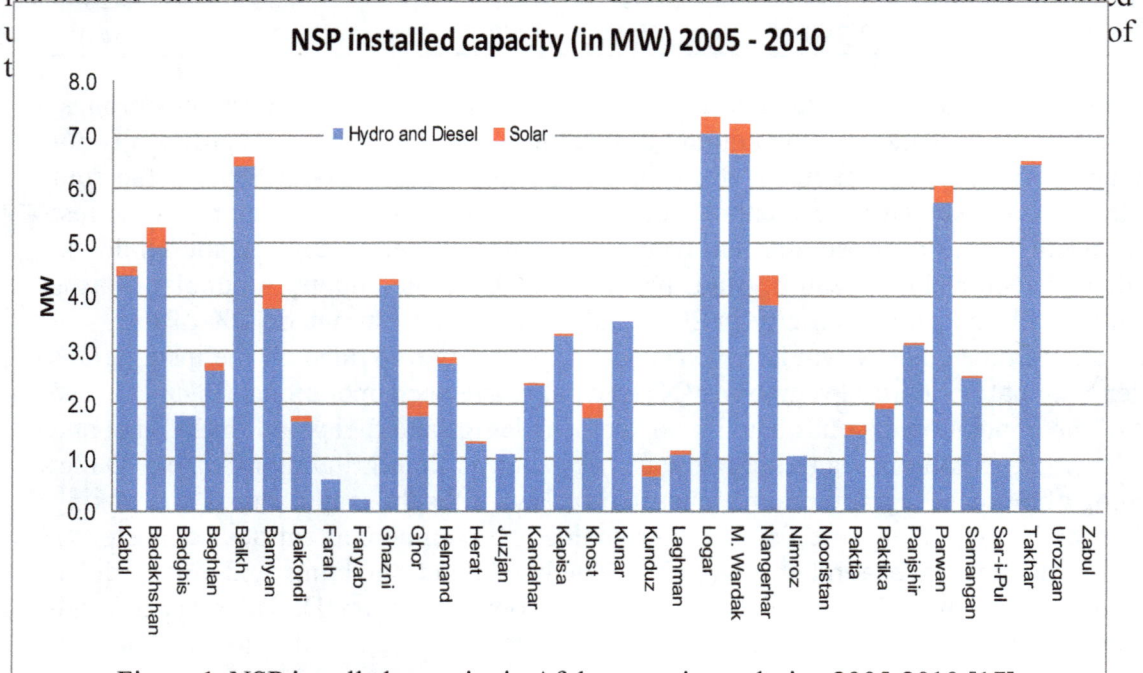

Figure 1. NSP installed capacity in Afghan provinces during 2005-2010 [17]

The total installed capacity under NSP between 2005 and 2010 was reported to be about 93 MW, which consist of about 28 MW from hydropower, about 58 MW from diesel and about 7 MW from solar energy [17]. The villagers have reported that many of the diesel generators and solar home systems installed under NSP are no longer operational. The authors have visited some SHS installations and found that most of the systems were out of use after two years from installation. The lead acid battery was not anymore functional and there was no battery replacement mechanism in place. High diesel prices, faulty installations and no equipment replacements (diesel engine and solar batteries) are the most common reasons cited. Also some of the MHP plants are not in proper operation. Estimate from different stakeholders lies between 20 and 40% (e.g. [18]) for non operational hydro capacity.

Besides the decentralized power supply projects, there is a transmission line project under implementation. The North East Power Supply (NEPS) program, a major transmission grid project in the country, is under construction to provide power to the north-eastern portion of Afghanistan, including Kabul. The infrastructure will help to import power from neighboring Tajikistan, Turkmenistan and Uzbekistan at significantly lower cost than diesel powered sources of electricity. For the isolated regions, the government further plans to implement off-grid systems based on renewable energy as mentioned in its national development strategy.

The main objective of this paper is to identify the status of MHP plants that were installed mainly under NSP in four North-eastern provinces of the country: Badakhshan, Baghlan, Balkh and Takhar (marked with solid-red borders in Figure 2) by carrying out the field monitoring of individual sites and to assess their socio economic impacts. Those survey outcomes mainly include – numbers of surveyed MHPs in individual province, MHP plants installation years, capacity range of individual MHP plant, turbine types, initial investment cost in USD/kW, project implementing agency (donor/private/government), daily MHP plant operation schedule, end user tariffs, productive use of the electricity, and the end user's satisfaction to the electricity supply. Results obtained thus, together with the information about other means of electricity generation and supply systems in these provinces and in a country as a whole, are expected to be an important dataset in order to develop a provincial and national electrification plan for different stakeholders.

Figure 2. Four studied provinces in the North-eastern Afghanistan (enclosed by solid border)
Source: www.mapsoftheworld.com

FIELD MONITORING

This paper limits its scope of analysis only in these four provinces mentioned above and only to the MHP plants. The questionnaires were developed by the authors or under the direct supervision of the authors. For the survey, qualified local electrical engineers were hired and sent to the villages with necessary knowledge and equipment. They were trained on the data collection and field monitoring procedure before they went to the fields. Authors of this paper could not visit the individual MHP installation sites themselves (except a few sites located nearby from the provincial capitals) due to local political and security concerns. However, they were in contact with the surveyors via phone whenever communication was necessary to ensure the accuracy and quality of the gathered information.

The survey questionnaire comprised altogether 47 points. Selected major points in the questionnaire included: date of survey; name of project site, village, district and province; name of CDC responsible for plant operation; number of families/houses/people supplied with the electricity; start and end year of plant construction; available water flow in the river; condition of civil infrastructure such as weir, forebay tank, spillway, settling basin, flushing gate, etc.; manufacturer, type, nominal capacity and number of turbines at the site; generator type and its rated capacity; initial project costs and the project implementing agency; plant operation hours in a day; operation model and operator's salary; electricity tariffs; productive use of electricity; end user's overall satisfaction with the electricity supply; and additional open comments from the informants. In most of the surveyed sites, photographs of river and canal intake, other civil components and powerhouse components (turbines, generators, etc.) were taken by the surveyors to support the data analysis afterwards.

Study area

This section gives a short overview of the four provinces studied. Badakhshan province is bordered by Takhar province in the west and Nuristan in the south (Figure 2). It shares international borders with Tajikistan in the north, China in the east, and Pakistan in the southeast. The province has 28 administrative districts. The province is home to about 860,300 people [8] and the provincial capital is Faizabad. The national electricity transmission and distribution grid does not link this province. Altogether 173 MHP plants distributed in 24 districts (out of 28 districts) of this province were visited during 2008-2010.

Baghlan province is bordered by five provinces: Parwan, Bamyan, Samangan, Kunduz and Takhar (Figure 2). It is divided into 15 administrative districts. The provincial capital is Pul-i-Khumri. The province is home to about 818,600 people [8]. The national grid passes through this province and the provincial capital has access to electricity. However, only about 9.5% of the province's population has access to the grid electricity [10]. In this province, the field survey was carried out in 2010 in its 5 districts and altogether 58 MHP plants were visited.

Balkh province is bordered by four provinces: Kunduz, Samangan, Sar-i-Pul and Jawzjan (Figure 2). It is divided into 15 administrative districts. The provincial capital is Mazar-i-Sharif. The province is home to about 1,169.000 people [8]. The national grid passes through this province and the provincial capital has access to grid electricity. In 2010, about 46% of the province's population was said to be supplied with grid electricity [10]. The survey was carried out in 2010 in 3 districts and altogether 15 sites were visited.

Takhar province has 17 administrative districts and its provincial capital is Taloqan. The province is home to about 886,400 people [8]. The national electricity grid does not link this province so far. The survey was carried out in 2009 in 11 districts and altogether 175 sites were visited. About 1.3% of the province's population living in Taloqan was supplied with diesel electricity from DABS in 2010. The electricity tariff was 66 ¢/kWh for residential customers (USD 1 = ~ AFN 53, Afghan afghani) [19]. This figure is quite higher than the DABS national grid average price of 11 ¢/kWh for residential customers, e.g. in Kabul.

RESULTS

The monitoring results on the MHP plants installed in the studied four provinces are discussed in this section. Not all the MHP plants installed in those provinces had been visited because of different reasons (e.g. security, transportation, time constraints, etc.). The summary of the surveyed sites is given in Table 3.

Table 3. Summary of MHPs survey results in four province

Description	Unit	Value
Total number of surveyed MHP plants	[Number]	421
Total installed capacity (considered only in 406 sites)	[MW]	10.7
Total operational capacity (out of 406 sites)	[MW]	7.6
Average cost of installation (excluding Badakhshan)	[USD/kW]	2,291
Total population to be supplied in 4 provinces	[People]	380,209 (~10%)

As can be calculated from Table 3, about 71% of the installed capacity was operational in those four provinces. The MHP plants in only 332 sites were found in proper operational condition during the time of field visits. Among the non-operational sites, about 38 sites were under construction during the time of survey. For the rest of the sites, either no information was available or they were not operational due to some defects in their civil, mechanical or electrical components. The classified results obtained and analyzed from the field monitoring surveys of these MHP plants are described in the following sub-sections.

Number of MHP plants surveyed

Figure 3 shows the MHP plants installation in individual province. In Badakhshan, the MHP installations after 2009 are not included. In the other three provinces, the shown numbers include the installations until 2010. These numbers also include MHP plants, which were under construction during the time of the visit. In Baghlan, although initial plan was to visit about 108 sites in 10 districts (out of total 15 districts of the province), only 58 sites distributed in five districts could be visited, due to the volatile security concerns in this province compared to other three provinces. In Balkh, altogether 15 sites were visited in 3 districts (out of 15 districts). In comparison, the number of installed MHP plants is relatively small in Balkh, as the grid electricity is available in provincial capital and some other districts. In Takhar, 11 districts (out of 17) of the provinces were visited.

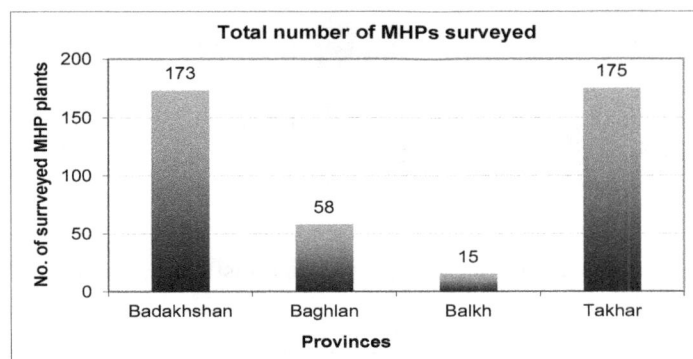

Figure 3. Surveyed numbers of MHP plants in each provinces

Installation years

Figure 4 shows the MHP plant installations over different years. Despite the political instability, recent years have seen a higher growth rate for new installations. The higher numbers after 2005 are mainly due to the donors supported NSP. There was no correlation between the plant installation year and the operation condition of the plants, i.e. also the older plants were performing well, while in some cases the newer plants were out of operation.

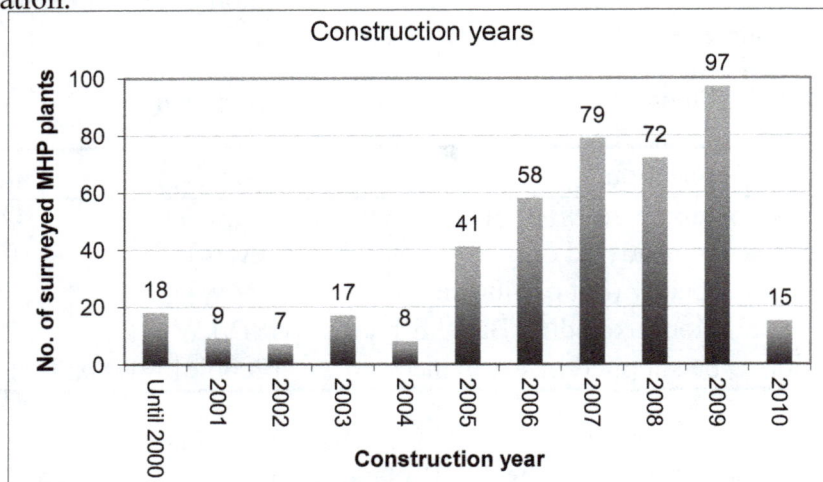

Figure 4. MHP plant installations over the years

Ranges of installed capacity

The survey team has collected data on the design capacity of individual power plants based on design reports. Most of the power plants operate during the night time, while the survey team visited the sites during the day time. This is why no immediate measurement of power output on the many sites could be made during the visit. It is difficult to say about the actual performance of these plants based on only information on nominal power output because the actual power output might differ largely in some cases. However, in this paper, the nominal design capacity of the turbines has been used in order to calculate the total ... (nominal turbine capacity a ...

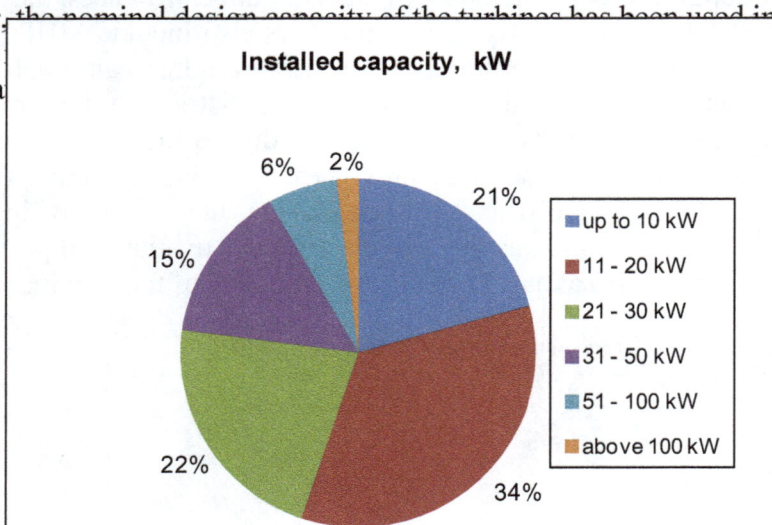

Figure 5. MHP plants according to their nominal installed capacity in kW

In Badakhshan, altogether 173 MHP plants were surveyed with a total installed capacity of 5.18 MW connecting to about 15% of the province's population. From the total installed capacity, only 57% of capacity (2.94 MW) was operational. In Baghlan, the total installed capacity has been calculated to be about 810 kW (in 58 sites). The plant capacity at individual sites varied from 5 to 30 kW. In Balkh, the total installed capacity has been calculated to be only about 299 kW (in 15 sites), while the plant capacity at individual sites varied from 10 to 48 kW. However, the total installed capacity of the operational power plants was only about 244 kW (in 13 sites). In Takhar, a total installed capacity of about 4.4 MW (in 160 sites) was reported, where only about 3.9 MW (in 144 sites) was operational. The nominal capacity at individual sites varied from 4 to 152 kW.

Turbine types

A large number of water wheels (82) are installed in Badakhshan. Additionally, there are 86 cross flow turbines, one Pelton turbine and three Francis turbines. In Baghlan and Balkh, all the sites that are operational use cross flow turbines. In Takhar, two sites use Francis turbines, 27 sites use water wheels and the majority of the sites (i.e. 129) are equipped with cross flow turbines, a popular turbine for small MHP plants in many other countries too. No information on turbine types was provided for the rest 17 sites in Takhar, stating some are under construction and no reason was mentioned for the others. One of the surveyed power plants (in Farkar district) had three cross flow turbines with a total installed capacity of about 152 kW (individual turbines of 54, 54 and 44 kW, respectively). Another 15 power plants had two cross flow turbines at each site and the remaining plants had only one turbine per site. Most of the cross flow turbines were manufactured within the country, mainly by the manufacturers located in Kabul. Figure 6 shows the turbine types. The term unknown in the figure refers to those MHP plants that were under construction.

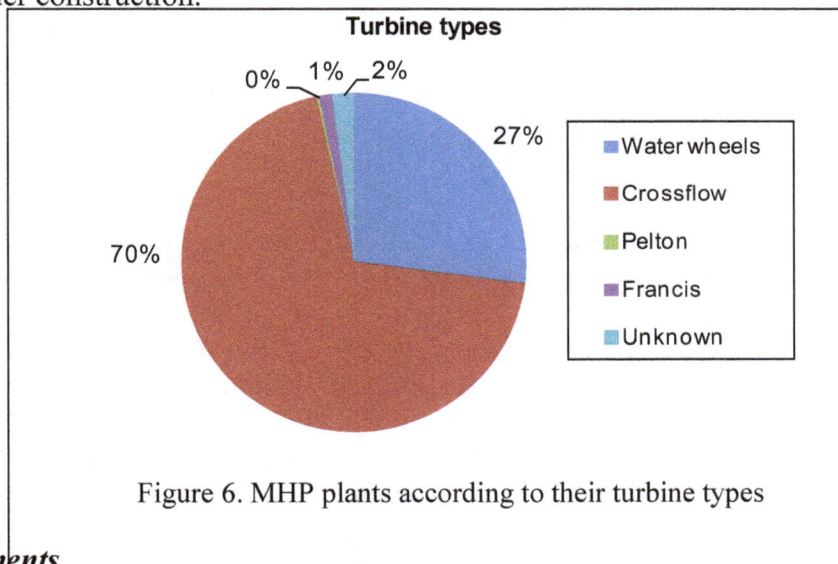

Figure 6. MHP plants according to their turbine types

Investments

Figure 7 shows the investment costs per unit installed capacity calculated for 215 surveyed sites in the provinces of Baghlan, Balkh and Takhar. No investment cost information was recorded for Badakhshan province. In Baghlan, about 83% of the installed capacity was more expensive than 2,500 USD/kW, which is far above the investment costs for MHP plants in Takhar province. The reasons were most likely the volatile security situation compared to other studied provinces. In Balkh, about 34% of the installed capacity was cheaper than 2,000 USD/kW and about 74% of the capacity was below 3,000 USD/kW. In Takhar, about 46% of the installed capacity was cheaper than 1,000 USD/kW, which is far lower than the investment costs for other renewable energy sources e.g. solar PV.

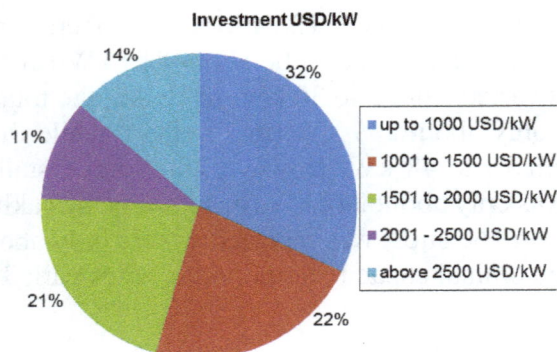

Figure 7. Share of MHP plants based on their initial investment cost in USD/kW

Project implementation agency

The majority of the projects were implemented jointly by community and NSP. In those sites, the community members basically support in construction work rather than cash payment. As a general rule, about 10% of the total investment cost is contributed by the community in NSP implemented projects. In Badakhshan, 81 units were installed under NSP, whereas as many as 90 units (mostly water wheels) were installed under private initiatives. Two of the units were implemented by DABS, each in Faizabad and Baharak. In Baghlan and Balkh, all the installations were under the NSP. In these two provinces, none of the units was installed under private initiatives unlike in Takhar and Badakhshan provinces. In Takhar, about 88% units were implemented under NSP. About 12% of the units were installed under private initiatives. One unit in Khwaja Ghar district (one cross flow turbine with an installed capacity of 36 kW) was implemented by the energy program of then German Technical Cooperation (GTZ, present GIZ). Aggregated in four provinces, about 73% of the projects were implemented under NSP, 26% under private initiatives and only 1% under DABS.

Plants operating time

Figure 8 to 11 show the daily operating schedule for the surveyed power plants in all four provinces. In Badakhshan, only two plants were reported to be operational round the clock. The majority of the plants operate only during the night. In Baghlan, among 42 reported plants, almost all were operational throughout the night. Also in Balkh, among reported 15 units, most of the plants were operational throughout the night. In Takhar, almost 47% (out of 153 reported) plants were operational throughout the night. Only two units - one operated by private sector and the other by community - were generating electricity round the clock. Both of them were supplying electricity also for productive uses.

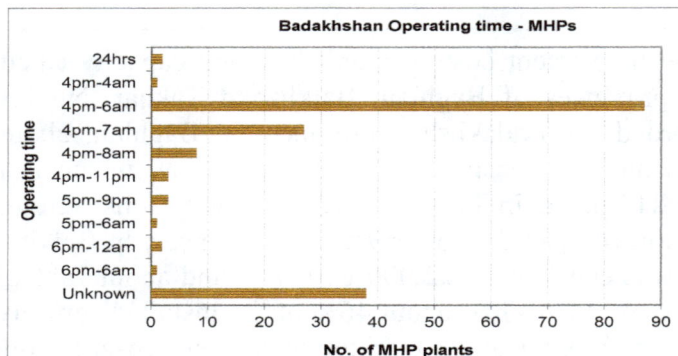

Figure 8. MHP plants operation time – Badakhshan

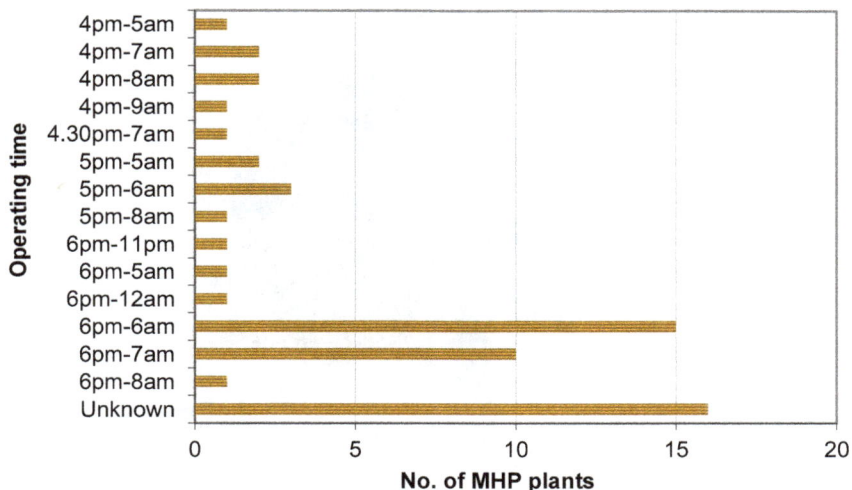

Figure 9. MHP plants operation time – Baghlan

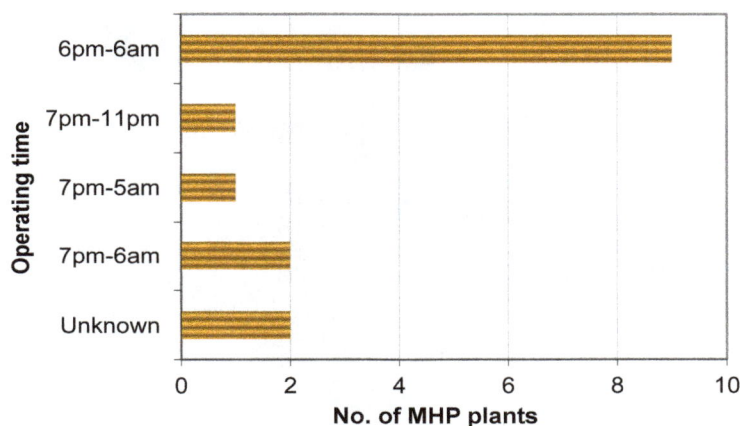

Figure 10. MHP plants operation time – Balkh

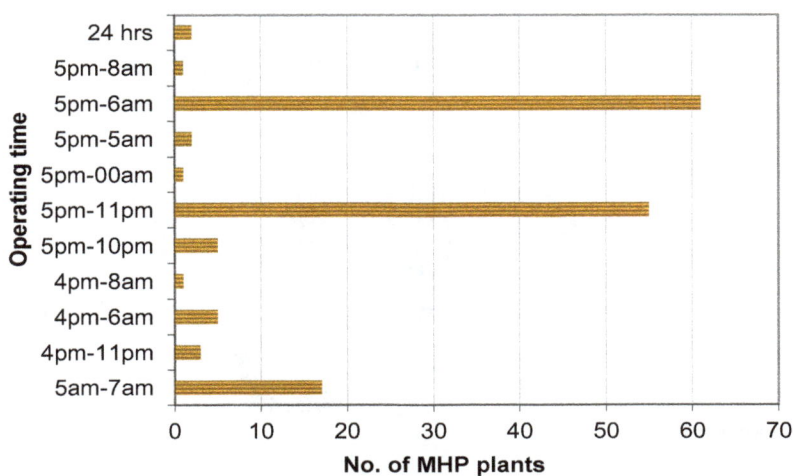

Figure 11. MHP plants operation time - Takhar

End user tariff

Figure 12 to 15 shows the end user electricity tariff (only for MHP plants) in four provinces.

Badakh Fee

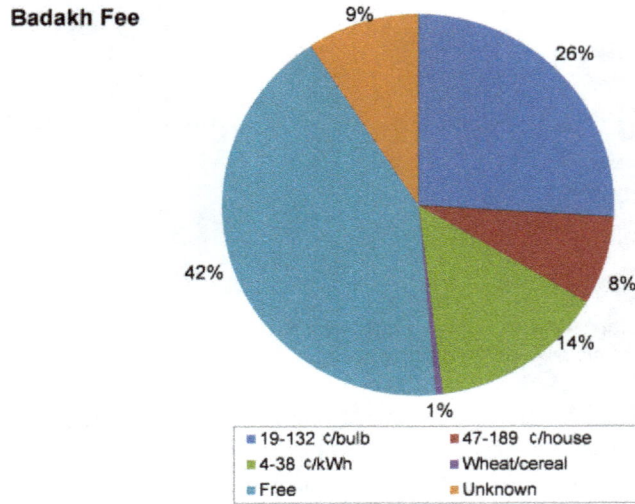

Figure 12. Electricity tariff for end users – Badakhshan

Baghlan Tariff

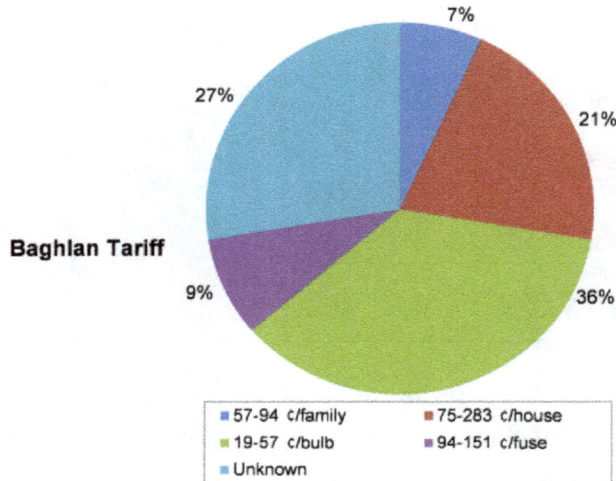

Figure 13. Electricity tariff for end users – Baghlan

Balkh Tariff

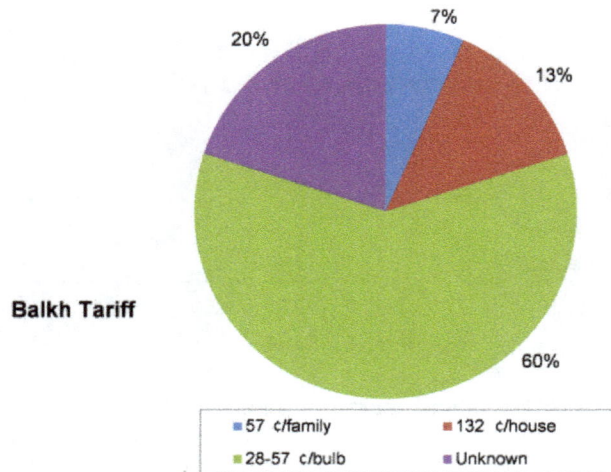

Figure 14. Electricity tariff for end users – Balkh

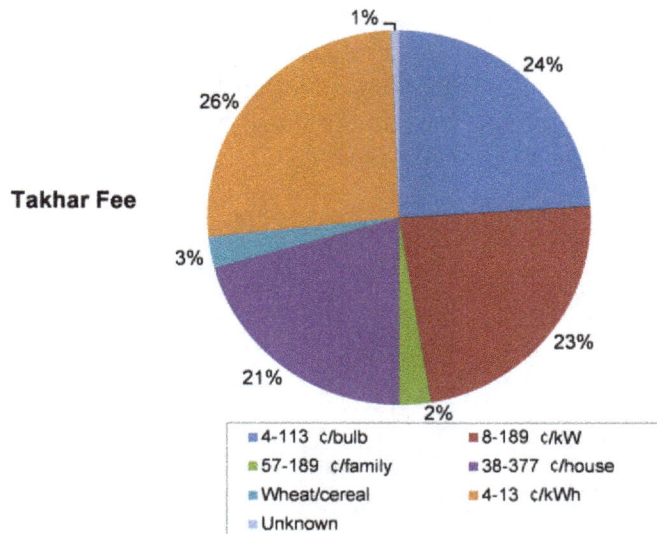

Figure 15. Electricity tariff for end users - Takhar

In Badakhshan, as many as 73 sites, mostly water wheels, offered the end users electricity or mechanical services free of cost. This certainly leads to problems with repair maintenance and thus it could be one of the reasons why the share of non-operational plants in this province is the highest. Other tariff structures were based on per bulb, or per house, or per kWh, etc. as shown in Figure 12. In Baghlan, the electricity generated from these power plants was not sold to the consumers based on meter reading. Instead, different flat rate tariff structures have been adopted, such as per bulb, per house, per family, per fuse (with power consumption limit of 200 Watt), etc. A range of fee structure for Baghlan can be seen in Figure 13. In Balkh too, different flat rate fee structures had been adopted as given in Figure 14. In Takhar, the electricity generated from about one fourth of the power plants (out of 150 sites for which such information is available) was sold to the consumers based on meter reading (i.e. per kWh basis). In the remaining plants, different flat rate tariff structures had been adopted as can be seen in Figure 15.

Productive use

Productive use means utilizing the electrical (and/or mechanical) energy for promoting employment, income generation, adding value to marketable products or services, etc. The productive use strategy may include a number of elements: awareness raising at target group, business idea generation and training of potential or existing business persons, access to micro finance, technical and vocational training, marketing support, etc. Such a strategy needs to be supplemented by studies about socio-economic potentials for energy planning and supply in the regional scale [20]. In this section, the major productive uses in survey sites refer to flour mills, oil pressing mills, saw mill, metal workshop and carpentry. Figure 16 shows the share of MHP plants (in number of units) supplying electricity for productive use.

In Badakhshan, only three units were supplying electricity for productive use. At the rest of the sites, the electricity generated from these plants had been exclusively used for household consumption and none of these plants supplied electricity for machinery and other appliances that are used for income generation. In Baghlan and Balkh, none of the units contributed to productive use. One of the reasons behind it could be relatively small capacities of these power plants. In Takhar, each MHP plant was supplying electricity to a single machine that is used for entrepreneurial use in about 38 sites and each plant was supplying electricity to two machines used for productive use in another four sites. Also the power plants in about 11 additional sites were reported to have been supplying the electricity to productive users, though no elaboration has been made. No such information was recorded for about 15% of the sites.

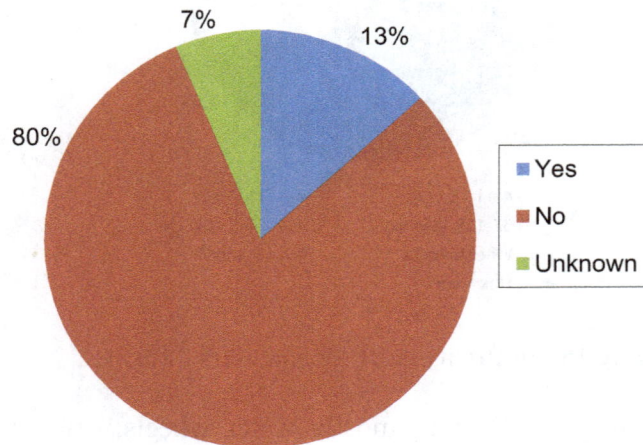

Figure 16. MHP plants supplying electricity also for productive use

Community satisfaction

By community satisfaction here is meant whether the end users get scheduled electricity supply without acute disruptions. In Badakhshan, the end users connected to 145 MHP plants were reported to be satisfied with the electricity supply. Those numbers for Baghlan and Balkh were at 41 and 10 sites, respectively. In Takhar, the end users connected to about 137 of the MHP units are reported to be satisfied with the use of electricity. The combined values for all four provinces show that the end users at about 79% of the sites were said to be satisfied, whereas 11% were said to be dissatisfied. For 10% of the sites, no response had been reported. Many reasons were mentioned for the cases of non-satisfaction, such as too many users were connected to a single supply system and the supply voltage was too low, power plants were not operating due to damage of certain parts, etc. Similarly, other reported common problems were leakage in canals, problems in intake and weir that were not made of concrete, etc. In some sites, during the flood season, intakes and weirs were swept away and villagers were without electricity for some days or weeks. However, it was also reported that the villagers repair such damages on their own; and therefore, the power plants could be operational again.

SUMMARY

The discussed survey gave an overview on the current status of the installed MHP plants in the four provinces studied. The following summary has been drawn from the monitoring:

- Almost 80% of the Afghan population still lack access to grid electricity supply. Decentralized power plants, mainly those based on hydro and solar resources, could play a vital role to supply these people with electricity;
- A significantly large number of rural electrification projects have been installed in the country, especially after 2005 under donor supported NSP programme. These projects, if all were operational, would supply the electricity needs of country's 7% population;

- In the four provinces studied, the recent MHP plants installation is in increasing trend. The choice of installation sites is not necessarily based on the optimal match of the local demand for electricity and the on-site resources availability, sometimes they are based on other preferences (donor's interests, security issues, etc.). This is why relatively safe Badakhshan and Takhar provinces see significantly higher number of installations. Comparably easier access to grid in Balkh and Baghlan provinces are other reasons for fewer installations there;

- Most of the installed MHP units have nominal capacities below 50 kW. Those small plants are mainly aimed to provide light to the end users. Therefore the consumption of electricity from such plants is relatively low and thereby less revenue. These plants may pose a financial problem once the plants need repair-maintenance work;

- Cross flow turbines are extensively used. The positive side of these types is that the domestic manufacturing (in Kabul) of these units ensures the repair-maintenance know-how in case of defects. Other countries, especially Nepal, have examples of technical success of installed MHP plants when the plant components are locally manufactured and installed;

- Investment cost for the installed plants varies in a large range, i.e. below 1,000 to above 2,500 USD/kW. The lower costs are difficult to justify; most likely they do not include the non-monetary community contribution in their calculation. The higher costs include extra expenses for transportation of the materials up to the project site and sometimes also security related expenses. The capacity costs for MHP plants in other countries are also in similar higher ranges, e.g. about 2,500-3,000 USD/kW for the plants in range of 50 kW [21]. The investment cost in the surveyed sites is mostly from the donors in the form of donations. For the sustainability of donor supported rural electrification projects, the end user's focus should be thus to cover the operational and maintenance costs;

- Most of the projects are implemented under NSP. Such projects are operated by the communities once the project is commissioned. The national authority responsible for the electricity supply in the country, DABS, has not been involved in those small projects;

- Most of the plants are not operated round the clock, as there is no demand for electricity. Most of the plants supply electricity for lighting. The revenue from the power plants depends on the electricity sales, thus idling the plants most of the time is not economically sound. Additional entrepreneurial demand should be created for the electricity to ensure the necessary repair maintenance of the plants whenever necessary as well as to boost the local economy;

- Different end user tariffs have been set, from flat rate basis to consumption basis (meter reading). Non monetary tariffs are also common, adjusted to the local socio-economic situation of the end users. One of the reasons for relatively small share of meter based tariffs is also due to additional meter costs. Once the demand for electricity could be increased, the use of meters could be promoted easier and it makes economically feasible too. Tariffs vary widely, both below and above the national average price for grid electricity of 11 ¢/kWh;

- The productive use of hydro-electricity is still minimal. One of the reasons was NSP's focus to promote rural electrification mainly for lighting purpose. Very small generation capacity of the majority of the power plants as well as single-phase under-voltage power supply could be some of the other reasons behind it. The awareness campaigns on the productive use of electricity as well as the support for the establishment of marketing mechanism of locally produced goods could help to increase the productive users in the future;

- The portion of the end user that is not satisfied with the supply of electricity is relatively low. This dissatisfaction could be largely eliminated if the operators at the respective sites are provided with a basic training about the repair maintenance of the MHP plants. This could at least help to repair some of the defects such as water leakage, small defects in generators and motors, etc. leading to the smooth operation of those plants,

- Depending upon the power generation potential at the specific site, as well as the electricity demand in its vicinity, some of the power plants could be upgraded to bigger sizes. The installation of larger capacity plants could reduce the initial investment cost per kW capacity. However, the management of such plants should be done in a professional way securing reliable and high quality power supply necessary for productive use. The realization of such projects needs further studies on individual site. For this, the first steps would be the identification of potential sites where the power output can be increased and the estimation of the rehabilitation costs. Although it depends on the surplus electricity generation at individual power plants as well as the period (hours of the day) of consumption in different villages, a study on the possibility of micro grid establishment in the province could be useful. Such a grid would not only maximize the electricity utilization as a whole, but also help to improve the power quality (better voltage stability). However, economic and political aspects of such micro grid formation should be analysed for the sustainability of such projects;

- There is no doubt that the end users in those Afghan villages are enjoying the common benefits of electrification similar to elsewhere: electricity for lighting, for communication devices (mobile charging, television, radio, etc.), and sometimes for other appliances and machinery (ironing, rice cooker, oil extraction machine operation, grains grinding, etc.). Other benefits include the village-street lighting, children's home study in the evening, less/no health hazards by avoiding the use of kerosene lamps, etc. However, under the scope of this study, these benefits are not surveyed and quantified.

CONCLUSIONS

In Badakhshan, 173 MHP plants were surveyed with a total installed capacity of 5.18 MW connecting about 15% of the total population in the province. The majority of MHPs (165) operate with capacities of far less than 100 kW. From the total installed capacity, only 57% of capacity (2.94 MW) was operational during the time of the survey. In Baghlan, the total installed capacity of the operational power plants was about 585 kW (in 40 sites) and about 225 kW was under construction (in 16 sites). The capacity in individual site varied from 5 to 30 kW. The investment cost for these power plants was quite high, i.e. about 83% of the installed capacity was more expensive than 2,500 USD/kW. In Balkh, the total installed capacity of the operational power plants was about 244 kW (in 13 sites). The capacity in individual site varied from 10 to 48 kW. In Takhar, if almost all of these MHPs would generate electricity to their design capacity, only about 17% of the province's population could have access to the electricity. The total installed capacity of MHP plants for 160 sites was about 4.4 MW. The installed capacity for individual plants varied between 4 and 152 kW.

The non-operating capacities in different provinces varied; the highest being in Badakhshan province, i.e. 43%. The same rate for Takhar province was only 10%. For Baghlan and Balkh, the operational rate is close to 100%. Theoretically, the rehabilitation or only repair-maintenance of non-operating capacity should be possible to bring them back to their normal operation. However, individual site specific studies are necessary for exact recommendations. Upon the complete rehabilitation of non-operating power plants (i.e. assuming 100% of the installed capacity being operational) the number of people that could be supplied with MHP electricity in four provinces would be about 380,000 i.e., 10% of the total population of those four provinces.

Based on the summarized points above, we conclude that the rural electrification is important in Afghan villages, but it can contribute to the local socio-economic development effectively only when the use of electricity is expanded beyond mere lighting to entrepreneurial uses. This ensures also the techno-economical sustainability of rural electrification projects.

ACKNOWLEDGEMENT

Authors would like to thank GIZ ESRA (then GTZ ESRA) programme for making the MHP plants field survey data available for the analysis. Parts of the results (only for Takhar province) were published in a GTZ ESRA internal report, which was written by the corresponding author of this paper.

REFERENCES

1. Aqeel, A. and Butt, M. S., The Relationship between Energy Consumption and Economic Growth in Pakistan, *Asia-Pacific Development Journal*, Vol. 8, No. 2, pp 101-110, 2001.
2. Kraft, J. and Kraft, A., On the Relationship between Energy and GNP, *Journal of Energy and Development*, Vol. 3, pp 401-403, 1978.
3. Lee, C.-C. and Chang, C.-P., Energy Consumption and Economic Growth in Asian Economies: A more Comprehensive Analysis using Panel Data, *Resource and Energy Economics*, Vol. 30, No. 1, pp 50-65, 2008.

4. Nexant, Subsidizing Rural Electrification in South Asia: An Introductory Guide, USAID SARI Energy Programme, 2004.
5. AEIC (Afghan Energy Information Centre), National Status Report - Generation, AEIC: Kabul, 2010.
6. AEIC (Afghan Energy Information Centre), Annual Electricity Production Report – 2010, AEIC: Kabul, 2011.
7. AEIC (Afghan Energy Information Centre), Annual production report - 2011, AEIC: Kabul, Afghanistan, 2012.
8. CSO, Population Database for Afghanistan (2009-10), CSO: Kabul, 2010.
9. CIA, CIA World Fact Book - Afghanistan, https://www.cia.gov/library/publications/the-world-factbook/geos/af.html, [Accessed: April-2013]
10. DABS (Da Afghanestan Breshna Sherkat), DABS Customer Database 2010, DABS: Kabul, 2011.
11. GTZ ESRA, Energieprogramm Afghanistan - Internal Reports, GTZ ESRA: Kabul, Afghanistan, 2010.
12. Mainali, B. and Silveira, S., Financing Off-grid Rural Electrification: Country Case Nepal, *Energy*, Vol. 36, No. 4, pp 2194-2201, 2011.

13. Pigaht, M. and van der Plas, R. J., Innovative Private Micro-hydro Power development in Rwanda, *Energy Policy*, Vol. 37, No. 11, pp 4753-4760, 2009.

14. NSP (National Solidarity Programme), National Solidarity Programme - Afghanistan, http://www.nspafghanistan.org/, [Accessed: November-2012]

15. Shoaib, A., Rural development in Afghanistan: The Impact of Renewable Energy, Willy Brandt School of Public Policy, University of Erfurt: Erfurt, Germany, 2011.

16. NSP (National Solidarity Programme), *NSP Operational Manual*, http://www.nspafghanistan.org/default.aspx?sel=16, [Accessed: November-2012]

17. NSP (National Solidarity Programme), Database on NSP Implemented Projects - An Internal Report, NSP: Kabul, 2011.

18. Nexant, MHP assessment for USAID, Nexant/USAID, 2006.

19. Xe, Xe Currency Converter, http://www.xe.com/ucc/full/, [Accessed: May-2013]

20. GTZ ESRA, Afghanistan Rural Renewable Energy Strategy, GTZ ESRA: Kabul, Afghanistan, 2010.

21. Mainali, B. and Silveira, S., Alternative Pathways for Providing Access to Electricity in Developing Countries, *Renewable Energy*, Vol. 57, No. 0, pp 299-310, 2013.

5

Hybrid Decentralised Energy for Remote Communities: Case Studies and the Analysis of the Potential Integration of Rain Energy

*Ying Miao[1], Yu Jia[*2]*

[1]Centre of Development Studies, University of Cambridge, Cambridge, United Kingdom
[2]Department of Engineering, University of Cambridge, Cambridge, United Kingdom
e-mail: yj252@cam.ac.uk

ABSTRACT

For remote underdeveloped and sparsely populated regions, the use of national power grids to provide electricity can be both unsustainable and impractical. In recent years, decentralised renewable power has gained popularity, endowing social benefits to the local inhabitants through clean rural electrification. However, power reliability and system autonomy are often the primary technical concerns as current systems are largely single source reliant. Hybrid power systems that utilise multiple complementary renewables can help to reduce the dependency on conventional unclean options. A few selected case studies for both single source and hybrid power systems are reviewed, analysing critical success factors and evaluating existing difficulties. The additional integration of the novel rain-powered kinetic-to-electric generator technology to the existing hybrid model is analysed. As with development in general, there is no one-size-fits-all solution to bringing power to remote communities and the most sustainable solution should be found through analysing local resources, environmental conditions and maximising local involvement.

KEYWORDS

Hybrid power, Rural electrification, Decentralised, Rain energy harvesting.

INTRODUCTION

In face of global climate change and pressing concerns for energy security, it is now accepted that a sustainable energy system is of paramount importance in energy development. According to [1], a sustainable energy system needs to adhere to the principles of an energy hierarchy: the most desirable being energy conservation, followed by sustainable production (renewable energy), and lastly degenerative energy depletion (fossil fuels). Under such contexts, the efficacy of traditional power grids are increasingly contested due to their inefficiency at delivering electricity to rural areas that are underdeveloped in terms of infrastructure and often too sparsely populated for the installation of power grids to be cost-effective. Decentralised energy, on the other hand, are better fitted for the sustainable model, in its ability to reduce generation and distribution inefficiencies and encourage the introduction of renewable modes of energy production.

Decentralised energy systems, often referred to as standalone systems because they are not connected to the electric grid, have several important features. First, their operational capacity is matched to the demand of the region, which assumes top priority. These systems are therefore ideal for remote areas where the system is required to

---

function at low plant load factors. Second, these systems are likely to rely on renewable energy sources such as solar photovoltaic (PV), hence their exertion on local resources will be limited because of their small scale operation. On the other hand, it also means their operation is likely to be seasonal, depending on the availability of, for example, sunshine and rainfall. Consequently, the major downfall of decentralised energy systems is the extra battery and storage costs, in order to meet off-peak demands. Unlike grid-connected systems, standalone systems cannot feed excess energy back onto the grid, and they either have to be stored at extra cost or be wasted [2]. Because of this major drawback, the decision between grid-connected system and a standalone-decentralised system is often hedged on economic feasibility and load factors.

A wide variety of renewable sources have all been considered for decentralised energy generation, including but not limited to biomass, solar energy, wind, and hydroelectric energy generations. Among them, decentralised PV systems are the most popular as the worldwide Solar Home Systems (SHS) installations have expanded rapidly in the last two decades with continuing growth of demand in the developing countries of Asia, South America and Africa [3]. However, the efficacies of such systems are reliant on the availability of solar irradiation at the locality and cannot be realistically expected to satisfy local electrical demands at all times. Consequently, energy generation from decentralised sources are often complementary to conventional power generation technologies (in most cases, kerosene and diesel). Alternative methods of decentralised energy generation are often tried as separate entities to PV and thus encounter much of the same problems faced by PV alone, producing less than satisfactory results.

In order to ascertain higher proportions of clean energy in decentralised energy production, hybrid power systems, especially hybrid *renewable* power systems, should be given more consideration. By using case studies, this paper will highlight the limitations of PV systems as a standalone decentralised energy system in itself, and give a brief overview of current attempts at hybrid power systems in some regions. Further to the current renewable sources, this paper suggests the use of rain energy harvesting as a complement to hybrid systems, for certain applicable regions, and provides an initial theoretical analysis of this novel technique.

ESTABLISHED USES OF DECENTRALISED SYSTEMS

Single renewable source systems

Amazon, Brazil. It has been estimated that approximately 1 in 6 households in the Amazon region lacks electric lighting and an estimated two thirds of the households are in demand for rural electricity [4]. One of the major attempts at decentralised energy production and distribution in the area is the PRODEEM initiative (The Energy Development Program in States and Municipalities - *Programa de Desenvolvimento Energético de Estados e Municípios*) implemented by the Ministry of Mines and Energy, with the express purpose of the electrification of rural areas that are otherwise unconnected by the main electric grid. Over 6000 photovoltaic systems were installed until the suspension of the project in 2003 [4], but over 80% of the planned systems failed to materialise [5]. Nonetheless, the amount of power involved exceeded 5.2 MW at peak operation, which ranked it amongst one of the larger PV based rural electrification programs amongst the developing countries of the world at the time [6].

Upon evaluation, several intrinsic weaknesses can be found: first, photovoltaic systems were the primary option invested in, despite the limited number of clear sunny days (partially cloudy being the most frequent) and season-dependent abundance of solar irradiation [7, 8] in the area and the availability of other untapped environmental energy sources such the uniquely high levels of precipitation [9]. Secondly, nearly half of the systems were mislaid and a third reported operational failure soon after installation, severely undermining the success rate of the project. Lastly, all of the PRODEEM equipment were imported, hence national involvement were limited, providing very little technological knowledge to the local communities. By extension, lack of technologically competent personnel and difficulty in acquiring spare parts were part of the major problems faced by PRODEEM [4].

Following PRODEEM, three more projects were attempted in similar veins: PROEOLICA, PROINFA and PCH-HOM, which aimed at rural electrification through various technologies including wind, biomass and hydroelectric energy. Very few projects met their targeted output: PROEOLICA's planned installation target was 1050 MW and only a staggering 2.7% was actually achieved upon its termination and incorporation into PROFINA in 2004 [5]. While it appears PROFINA has met its deliverables, a significant portion of the reported power yield was borrowed from the relatively successful PCH-HOM projects. It is clear that the lack of coordination and evaluation between different programmes meant that advances are made sloppily and prior faults are not addressed properly before the objectives, targets and deliverables are mingled and substituted by other projects.

India. The Indian government has made several attempts at promoting the use of decentralised renewable energy, such as solar and biofuels, both in terms of legislations and execution. Integrated Energy Policy Report projects up to a third of the Indian energy mix might be made up of renewables by 2032 [10]. The pressing concern for clean energy in India is twofold: lack of electricity for lighting and appliances in some regions, as well as the use of traditional fuel that threatens the health and safety of rural Indians. However, the encouraged shift in consumption patterns toward energy produced from the bio-fuel plants have only shown moderate improvement in the overall effective energy efficiency [11]. Nevertheless, a four-fold increase in the annual growth rate of commercial establishments in the area was observed from 1996 to 1999 along with noticeable increase in job creations.

To complement sustainability, solar lanterns such as those used in Sagardeep Island also achieved a degree of success [12]. For effectively regulated demand and load conditions, solar energy acts as an economically incentive and technologically viable option in contrast to the installation of main power grids for this remote region. Again the social impact of electrification is evident: aside from the ability to carry on work in the night, some of the hospitals on Sagardeep now have 24 hour electricity supply and the availability of street lights has made the island safer. The employment of PV powered mini-grid system in the vicinity rather than standalone PV also delivers more stable mains AC power, which enables the operation of small electrical machinery for the village industries. The availability of electricity at night also encouraged locals to participate in night-time entertainment, fostering domestic industry and boosting local economy.

The main setback for solar is its seasonal availability: during monsoon seasons the power output is reduced to four hours due to cloud cover and the highest loss of load hours occurred [12]. Additionally, the ideal silicon-based PV operational condition is well known to be that of a relatively cold and sunny environment rather than the hot and humid climate of the region. It is clear that while the utilisation of solar energy is more sustainable, it will likely fall short to be the primary energy source in meeting the ever-rising rural demand.

Bangladesh. In 2010, 91% of domestic cooking in Bangladesh still relies on traditional 'unclean' biomass and around half of the population is without access to electricity [13]. As the average daily solar irradiation in Bangladesh amounts to 4 kWh/m^2 to 6.5 kWh/m^2 [14] during sunny months, compared to the global average of 3.61 kWh/m^2 to 7.96 kWh/m^2 [15], PV systems as a form of decentralised energy has encouraging potentials. As of 2013, an estimated 2 million systems have been installed in the country [16]. The critical success of these programmes owed much to the microcredit institution of Grameen Shakti, which sold PV systems on credit to rural households, as well as NGOs such as The Centre for Mass Education in Science and The Bangladesh Rural Advancement Committee, which promoted the technology [14]. Not only does the private sector, such as IDCOL, sponsor household SHS systems, but solar irrigation pump projects and mini-grid projects are also attempted to provide cleaner energy to communal buildings (schools, hospitals) and agricultural lands [17]. The joint efforts of both the public and private sector in the dissemination of PV systems are paramount in ensuring the success of the SHS system in Bangladesh.

However, Bangladesh shares similar seasonal monsoon and high temperature issues as its Indian neighbour of West Bengal. Hydroelectric power is severely limited in Bangladesh due to the flatness of the terrain and the contentious issue of downstream water sharing with India. Wind turbines are only applicable in coastal areas where strong winds immediately preceding and after the monsoon season may be harnessed to complement existing or other forms of generation [18].

Hybrid power systems

Current attempts at utilising hybrid power systems largely rely on a hybrid system of both renewable and non-renewable (conventional) energy generation: diesel power is often the choice complement to renewable energy systems because of its low initial price of installation and easy fuel storage. NREL's HOMER software (a renewable energy modelling software) has been a popular tool to estimate and aid real implementations in attempting to converge multiple energy sources into a more efficient mutually complementary system. A few combinations of such systems are outlined below:

Photovoltaic-diesel systems. A 50 kW PV-diesel hybrid system has been installed in the village of Campinas, of the scarcely and sparsely populated Amazonas state of the North region of Brazil since 1996, providing adequate power for a total of 120 households [19, 20]. A further PV system of 20 kW was added to the existing diesel generator of the village of Araras of the nearby Rondonia state in the same region of Brazil, in 2001. The economic feasibility were simulated [21] to be dependent on the increase of costs of diesel fuels: without subsidies, a 15% increase in diesel price means the substitution of 50 kW diesel system into hybrid becomes economical; while at a 45% increase in diesel price in some regions, 100 kW systems could be economically converted into hybrid systems.

Similarly, another simulation study [15] has proposed a potentially viable hybrid system of 4 kW PV system together with 10 kW diesel system and a battery storage of 3 hours of autonomy for use in hot regions such as Dhahran, Saudi Arabia. It was found that the percentage of fuel saving compared to diesel-only systems are 19%, with a total reduction of carbon emission up to 2 tons per year.

Photovoltaic-wind-diesel systems. The Tamaruteua village in Para State of North Brazil have utilitised the photovoltaic-wind-diesel system since 1999 and its power output was doubled when upgraded in 2005. The system features a PV system of 3.84 kW_p, a combined wind turbine power of 15 kW_p and a diesel generation of 36 kW [20].

Simulated results [22] suggest the combination of these three energy sources increase the overall reliability of the system, as well as reducing CO_2 emissions significantly. In the case of Algeria, Saheb-Koussa *et al.* has outlined the complementary nature of PV and wind systems, as days without sunshine often coincides with days with high wind velocity at the site. For sites tested, not only household demands are met but surplus energy are also available several months of the year, allowing extra energy expenditure in the area [23].

Wind-photovoltaic systems. The Joanes Village of Para state, North Brazil have employed a PV system of 10.2 kW_p and a total wind turbine generator standing at a capacity of 40 kW_p to act as a complementary system to the grid during consumption peak times, as well as a standalone system, to 170 families in the region [20]. A simulation-based study has demonstrated that under reasonable energy, battery and load conditions, hybrid power systems are able to noticeably enhance energy supply stability over single PV or wind systems [24].

Understandably, the main concerns underlining hybrid power systems, and decentralised power systems in general, are system reliability, its level of autonomy and the availability as well as the abundance of the specific sources of renewable energy at the chosen site. As a result, diversifying energy sources would increase system autonomy and cater for the instances where one source of renewable energy falls short, such as cloudy days or monsoon seasons for PV systems. It is with these implications in mind that this paper proposes the introduction of the relatively under-explored rain energy to be included in future designs for decentralised hybrid power systems.

RAIN ENERGY AS A COMPLEMENT TO HYBRID SYSTEMS

In regions with high levels of rainfall, harvesting raindrop energy can be an additional alternative option and can be employed to help promote the availability of decentralised energy for either small electronics such as remote and wireless telecommunication devices on an individual device-level, or the deployment of an array of rain energy harvesters to complement conventional decentralised renewable power generation solutions. The remote northern regions of the Brazilian Amazon and isolated islands of east India and Bangladesh are examples of potential candidates that receive an abundant level of seasonal rainfall while conventional infrastructural connectivity to the national power grid can be both challenging and uneconomical [13].

Figure 1 presents a few selected examples of most rain abundant developing regions of the world compared to other regions that receive relatively moderate amounts of rainfall [25].

Rain abundant developing regions

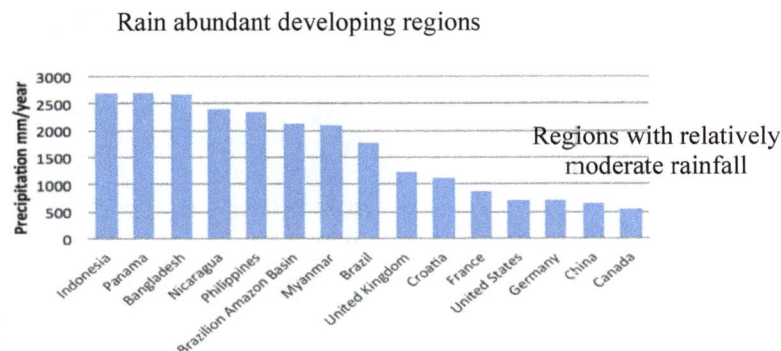

Figure 1. Precipitation data for a few selected regions (data based on [9, 25])

Figure 2 illustrates the level of electrification in these respective developing regions [26]. For instance, approximately half of the Bangladeshi population still lack access to the national grid. Potential employment of solar and wind sources to fill some of this gap are limited by seasonal dependency: solar irradiation peaking between March and May, and wind speeds peaking during June and August [27]. Furthermore, of the average annual rainfall of 1.35 trillion cubic metres that Bangladesh receives, 80% occurs during the Monsoon season between the months of June and October [18, 28]. Therefore, the addition of rain to the hybrid mix of renewables helps to promote the seasonal interchangeability and continuality of the various environmental energy sources.

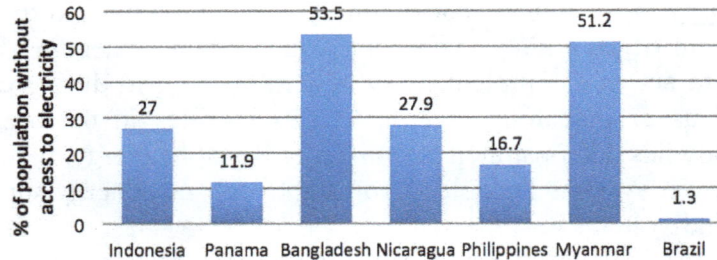

Figure 2. Electrification for selected rain abundant developing regions (2010 data based on [26])

Kinetic energy of the rain

As rain droplets fall through the atmosphere, kinetic energy builds up until the terminal velocity is attained. Upon impact, non-elastic collision takes place and the energy is released. Energy can be harvested either from the instantaneous impact of the individual droplets or through the collection of rainwater at an elevated platform and realising flow induced generation of the collected water mass. The later scenario is analogous to a mechanical winding mechanism where small mechanical energy can accumulated to a large kinetic release and parallels can be drawn from hydroelectric generation techniques.

Physics of raindrops

The precise shape of a falling raindrop primarily depends on its size. Assuming the absence of stochastic variables such the effect of wind, a spherical model can be adopted for a relatively small (≤ 2 mm in diameter) falling raindrop [29, 30]. This assumption derives from the surface tension of water, which holds the raindrop together against air drag. With increasing size, air pressure overcomes surface tension and deforms the droplet until a rip takes place to break it into smaller spherical droplets again [31]. Typically a single water drop larger than 6 mm does not survive the ripping effect of air drag during free fall on earth. For the purpose of this analysis, the spherical approximation is adopted as shown in Figure 3 and Equations 1-4 can be observed.

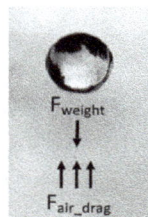

Figure 3. Forces acting on a free falling droplet, assuming spherical shape (accuracy diverges for diameter >2 mm)

$$F_{\text{weight}} = mg = \rho_{\text{water}} V g = \frac{\rho_{\text{water}} 4\pi r^3 g}{3} \tag{1}$$

where, F_{weight} is the gravitational force exerted on the raindrop, m is the mass of the raindrop and g is the acceleration due to gravity, ρ_{water} is the density of water (1000 kg/m^3 at 277.15 K for pure water), V is the volume of the raindrop and r is the radius of the raindrop.

$$F_{\text{air\_drag}} = \frac{\rho_{\text{air}} A C_d v^2}{2} = \frac{\rho_{\text{air}} \pi r^2 C_d v^2}{2} \tag{2}$$

$F_{\text{air\_drag}}$ is the resistive force exerted by the air pressure and ρ_{air} is the air density (1.225 kg/m^3 at 288.15 K and 1 atm), A is the cross-sectional surface area of the falling object, C_d is the drag coefficient (0.47 for a sphere) and v is the velocity. At terminal velocity v_t, F_{weight} equals $F_{\text{air\_drag}}$.

$$\frac{\rho_{\text{water}} 4\pi r^3 g}{3} = \frac{\rho_{\text{air}} \pi r^2 C_d v^2}{2}$$

(3)

$$v_t = \left(\frac{8\rho_{\text{water}} r g}{3\rho_{\text{air}} C_d}\right)^{\frac{1}{2}} \tag{4}$$

Terminal velocity under 1 atm is around 3 m/s for 1 mm droplets and approximately 10 m/s for 5 mm or larger droplets [32]. Kinetic energy is directly related to the squared of velocity.

Harvesting the kinetic energy of rainfall

The conversion of kinetic energy to electrical energy can be achieved via a number of transduction mechanisms, including electromagnetic, piezoelectric and electrostatic [33]. The few rare existing literature on harvesting rainfall energy has opted for piezoelectric material as the primary transducer [29, 30, 34]. Piezoelectric material yields an electrical charge polarisation when mechanical strain is induced, and vice versa. Popular piezoelectric material for energy harvesting include lead zirocnate titanate (PZT), polyvinylidene fluoride (PVDF) and aluminium nitride (AlN) in descending order of the piezoelectric strain constant, which can be measured in coulomb charge polarised per newton force applied. Therefore, upon direct kinetic impact onto these materials, electrical energy can be generated.

Assuming a mass-spring-damper model for the kinetic-to-electric energy conversion, Equation 5 can be observed.

$$m\ddot{x} + c\dot{x} + kx = -m\ddot{y}(t) \tag{5}$$

where, m is the seismic mass, c is the damping constant, k is the spring constant, x is the relative displacement, y is the excitation displacement.

$$y(t) = Y_0 \cos(\omega t) \tag{6}$$

where, Y_0 is the excitation force applied and ω is the forcing frequency. Total power dissipated $P(\omega)$ by the damper system is given by Equation 7.

$$P(\omega) = \frac{m\zeta Y_0^2 (\omega/\omega_n)^3 \omega^3}{[1-(\omega/\omega_n)^2 + [2\zeta(\omega/\omega_n)]^2}$$ (7)

where, ζ is the damping ratio (damping by critical damping) and ω_n is the natural frequency of the structure. Critical damping is given by $2\sqrt{mk}$. At resonant frequency $\omega_n = \sqrt{k/m}$, fundamental mode of direct resonant power P_r can be observed in Equation 8. Resonant response can act as a form of further mechanical amplification to promote the conversion efficiency.

$$P_r = \frac{mY_0^2 \omega_n^3}{4\zeta}$$ (8)

Electrical power is extracted through applying electrical damping, which in turn forms part of the total damping. Therefore, this power parameter contains both electrical power output and parasitic power dissipation. Electrical damping representing power output can be observed in Equation 9.

$$P_r = (\frac{\zeta_e}{\zeta_T^2})(\frac{mY_0^2 \omega_n^3}{4})$$ (9)

where, ζ_e is the electrical damping ratio and ζ_T is the total damping ratio. Maximum electrical power at resonance can be achieved when impedance matching equates electrical damping to the sum of parasitic mechanical damping. Further parameters not taken into account by viscous damping include forward and backward coupling between the mechanical and electrical domains during the kinetic-to-electric energy conversion cycle of the piezoelectric material, which will have a bearing on the overall damping factor.

Technique I: instantaneous impact-based generation

This relies on direct impact induced kinetic-to-electric energy transfer via piezoelectric material [29, 30]. This mechanism is illustrated in Figure 4.

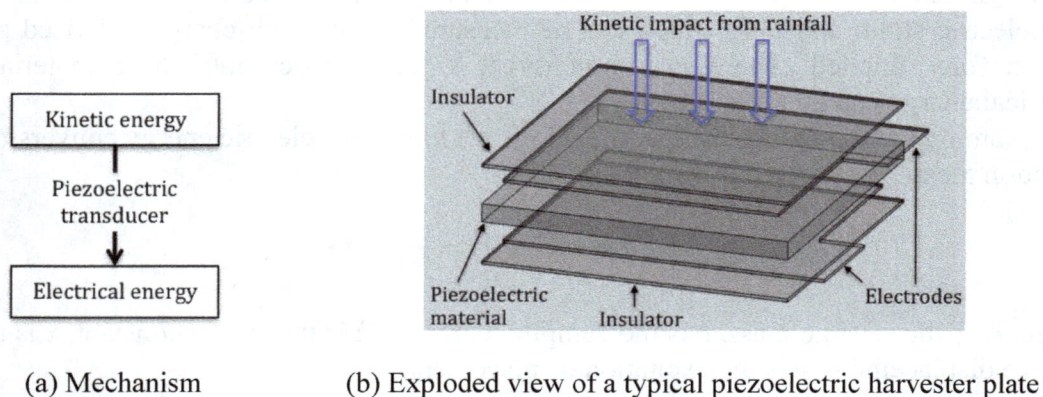

(a) Mechanism   (b) Exploded view of a typical piezoelectric harvester plate

Figure 4. Direct impact-based rainfall energy harvesting using piezoelectric transducers

The recoverable instantaneous power densities have been experimentally reported to range from 10^{-5} W/m^3 to 10 W/m^3 for small drizzles (mostly 1 mm droplets at 2.8 m/s) to heavy downpours (mostly 5 mm droplets at 5.7 m/s) respectively [30]. The several orders of magnitude higher energy density for rainfall with higher kinetic energy demonstrate the fast increasing energy conversion efficiency at higher excitation levels. This is partially due to a higher percentage of parasitic leakage from the transducers at lower piezoelectric polarisations.

Pros:

- The kinetic energy build-up is transferred directly to the transducer;
- Mechanism is simple and straightforward, which is important for mass production;
- Small on-board solution is possible, for integration with remote and wireless electronics such as self-sustaining sensor motes for monitoring and early-warning applications.

Cons:

- Individual impacts only release small energies. Conversion efficiency is poor for the small voltage output. Optimal power output is only achieved for large downpour of rain;
- Energy dissipation from inelastic collisions further reduces the conversion efficiency;
- Upon attaining terminal velocity, the kinetic energy of the raindrops no longer increases. Therefore, there is no difference in power output between harvesting rainfall on ground level or at a higher elevation where terminal velocity has already been attained.

Technique II: potential energy collection-based generation

A collection mechanism can be employed to build up a reservoir of water from rainfall catchment at a higher than ground level elevation. The gravitational energy accumulated over time can then be released and channelled through a fluidic flow mechanism to drive a kinetic-to-electric transducer. This enables a larger instantaneous driving force on the transducer than that achievable from direct impact, which allows electrical operation at a higher efficiency. The basic outline of the mechanism is shown in Figure 5.

Watermill or turbines are traditional designs for harnessing hydroelectric power. Both electromagnetic generators and piezoelectric plucking transducers can be employed at the core of these rotational generators. The inclusion of a reservoir acts as an energy buffer and allows a more continuous and less time-varying operation of the subsequent transduction mechanism. A piezoelectric watermill driven from a water tank has been estimated to yield average energy densities in the order of 10 W/m^3 to 100 W/m^3 [34].

Rotational generators require a certain flow rate to overcome the inertia and operate at an optimal speed. Water can also be channelled to enable a direct vibrational excitation on a piezoelectric generator [35, 36]. This linear alternative is less susceptible to the inertia issue and depends more on the pressure induced by the fluidic channels. The linear generator design iteration illustrated in Figure 5b operates with impact induced plucking of a cantilever beam, which can be electrically coupled to either a piezoelectric or electromagnetic transducer. This mechanism is similar to human motion harvesting from plucking of piezoelectric beams [37].

Pros:

- Large kinetic energy release is possible from the potential build-up to achieve higher electrical efficiency for the transducers;

- Conventional electromagnetic hydroelectric generation technology can be incorporated;
- Although the power efficiency of electromagnetism does not scale well downwards, piezoelectric generator turbines can be employed to retain the power efficiency at smaller decentralised scales.

Cons:

- The kinetic energy accumulated by the raindrops through its free fall is lost;
- Amount of energy releasable depends on elevation achievable;
- Large-scale mechanism is required to maximise rainfall catchment.

Both direct-impact and collection-based mechanisms can be simultaneously employed to complement each other. Rainwater following direct impact on piezoelectric surfaces can then be channelled to a collection reservoir.

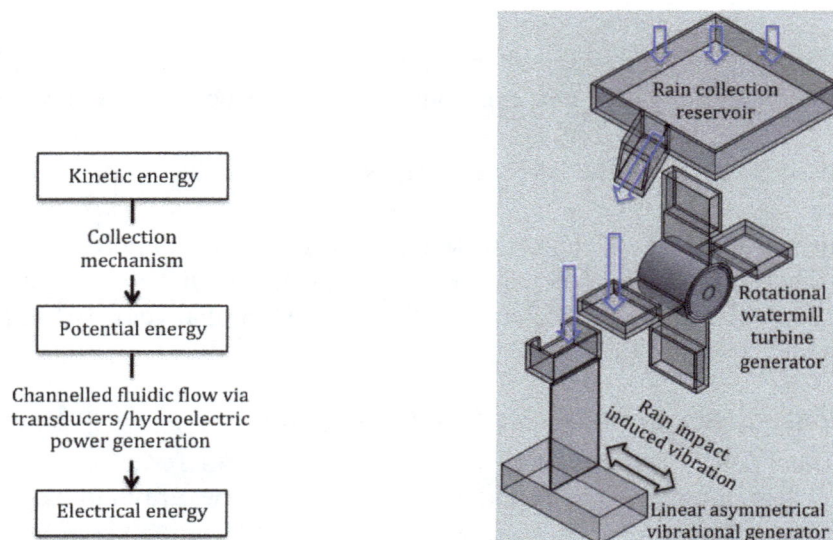

(a) Mechanism   (b) Design iterations of collection-based rain harvester

Figure 5. Potential energy collection-based rainfall energy harvesting. Release of reservoir as large kinetic energy can be fed to either linear or rotational kinetic-to-electric transducers

Simulated response of kinetic loading of a piezoelectric plate by rain droplets

Unlike the conventional electromagnetic generators that rely on higher displacement/velocity to maximise power output, piezoelectric material focuses on strain maximisation. Simulated COMSOL Multiphysics solid mechanics models are presented in this subsection to better understand the resultant strain from the kinetic loading of a piezoelectric plate from rainfall. The piezoelectric material chosen is the popular PZT-5H, which has a relatively high piezoelectric strain constant and the plate dimension is 50 mm by 50 mm and constrained on the four perimeter edges. A 5 mm diameter raindrop is assumed, which results in approximately 6.42×10^{-4} N of force and 32.7 N/m^2 of pressure assuming evenly distributed spherical raindrops.

Figures 6 and 7 show the displacement and first principal axis strain response from a single droplet loading at the centre of the plate and uniformly distributed loading by evenly spread droplets across the entire plate respectively. It can be seen that apart from straining the vicinity surrounding the loading point, there are also strain effects near the edges of the anchor. This effect is further amplified from the accumulated super-positioning of the evenly spread raindrops across the plate. Figure 7b illustrates strain concentration near both the centre and the edges. This is due to the maximisation of both bending strain near the more flexible plate centre and volumetric strain from the Poisson's effect near the less flexible anchored edges.

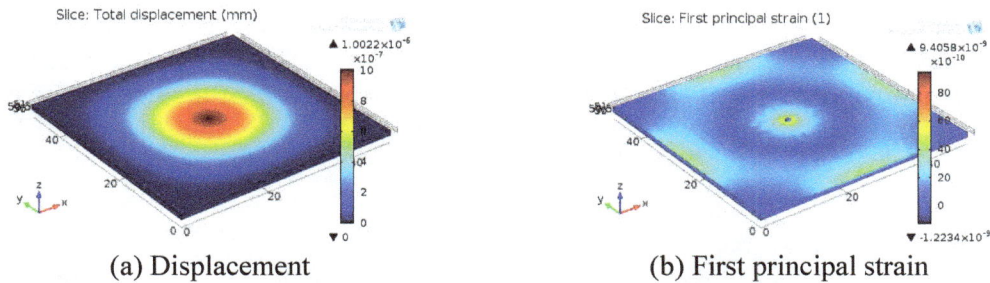

(a) Displacement

(b) First principal strain

Figure 6. Loading at the centre of the piezo plate by a single 5 mm rain droplet ($\sim 6.42 \times 10^{-4}$ N)

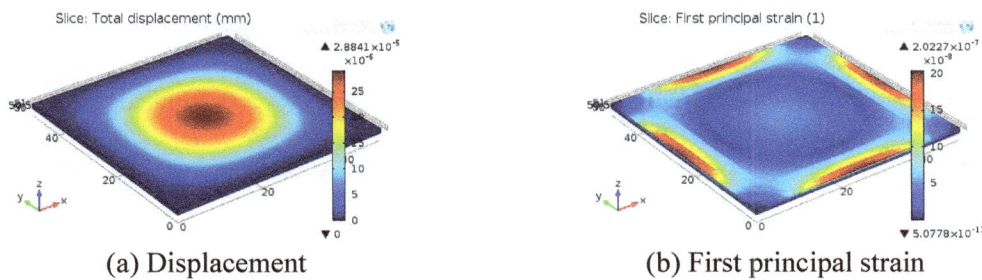

(a) Displacement

(b) First principal strain

Figure 7. Uniformly distributed loading on the piezo plate by 5 mm raindrops spread evenly over the plate (assuming 32.7 N/m^2 of pressure)

Figure 8 further presents the strain response from a single 5 mm diameter droplet loading at various locations on the plate and its respective effects on the anchor strain. It can be seen that the anchor strain is universal regardless of the position of the loading.

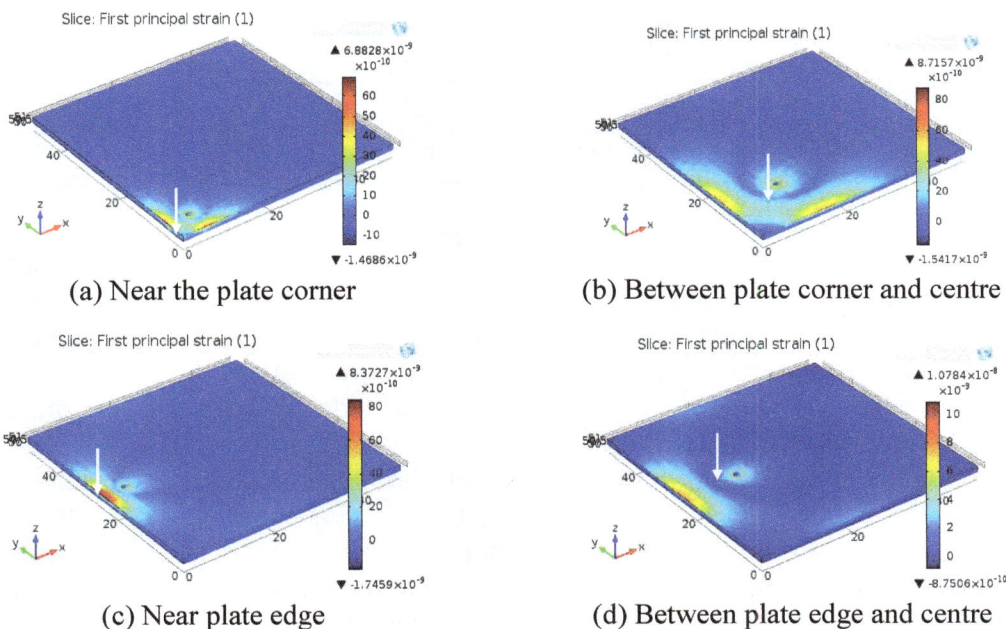

(a) Near the plate corner

(b) Between plate corner and centre

(c) Near plate edge

(d) Between plate edge and centre

Figure 8. Strain response (related to power output) from loading of the piezo plate by a single 5 mm rain droplet (assuming 6.42×10^{-4} N of force) at various locations off centre

Figure 9 illustrates the strain energy density achievable from the single droplet and evenly spread rainfall loading of the plate. The same accumulated edge strain effect can be observed for the distributed raindrop model, amounting to several orders of magnitude higher energy density over the single direct impact points. Typically, energy dissipation from anchor loss is problematic in many sensing applications, but in this scenario, energy can be harnessed near the anchors to exploit this phenomenon. Therefore, multiple arrays of small edge-anchored plates would fare significantly better than a single large plate anchored at far edges.

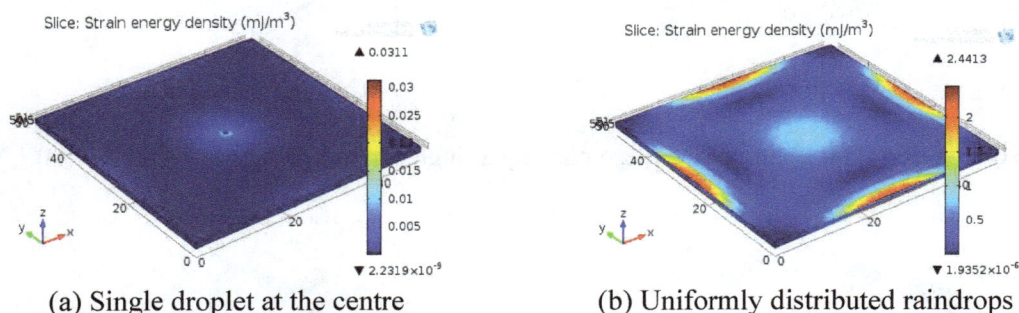

(a) Single droplet at the centre   (b) Uniformly distributed raindrops

Figure 9. Strain energy density induced from the kinetic loading of the piezo plate

DISCUSSION

Table 1 illustrates a basic comparison of power density achievable from rain harvesting reported in the previous section against typical power density values of solar and wind generators.

Table 1. Comparison of typical power density values between rain, solar and wind energy.

Source	Power density	Assumption
Rain	10 to 100 W/m^{3*}	1 to ~10 m/s rain velocity
Solar	10 to 100 W/m^{2**}	100 W/m^2 to ~1 kW/m^2 irradiance
Wind	10 to 100 W/m^{2***}	1 to ~10 m/s wind velocity

* Based on values estimated for potential energy collection-based generation.
**Values based on ~10% conversion efficiency. Factors such as angle of irradiation, unintentional shading as well as the size of the mounting frame further reduce the volumetric power density.
***Values based on power density = 0.5 × density of air × velocity^3. Inclusion of turbine size and required vertical elevation further reduce the overall volumetric power density.

It can be seen that rain energy fares reasonably and serves as a valid potential candidate as a part of the hybrid power mix. However, these are instantaneous power values under ideal loading conditions. Therefore, the accumulative and average power output for a specific region still requires a case study based analysis in order to compare the available power for each of the energy sources in order to determine the practical feasibility of adapting such a hybrid solution.

Figure 10 illustrates how different sources of alternative energy can be integrated together to offer a more robust hybrid system, either simultaneously or in an alternating fashion. Depending on the location or seasonal availability/abundance of a particular energy source, harnessing additional energy sources can help to promote the energy autonomy of the decentralised system. The development of smart power conditioning and power management subsystems are crucial to help to regulate and maintain the load-dependent peak power conversion and storage efficiencies. The generated power can either be directly used by individual devices and applications such as monitoring systems or fed back into a local mini electric grid or energy storage to contribute to the power supply of the local community.

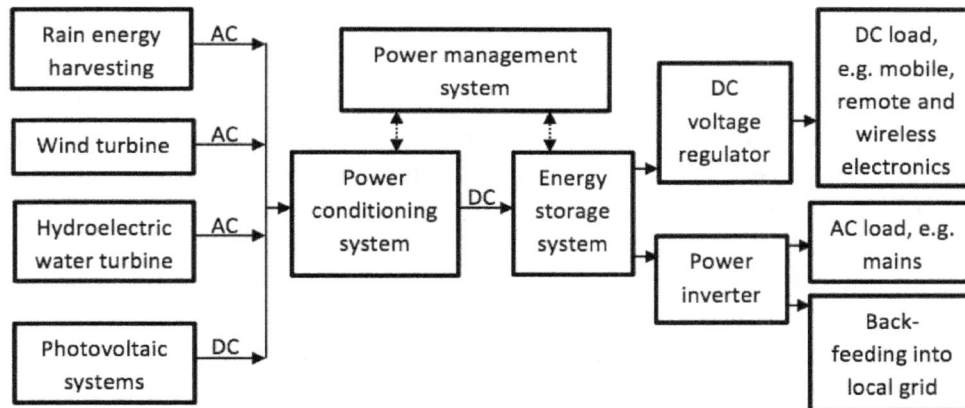

Figure 10. Integration of multiple complementary alternative energy generation technologies

Potential issues surrounding rural electrification can be divided largely into three categories: economic, policy and institutional [38]. High cost, lack of financing, and the lack of credit on the user end are the main economic concerns, whereas the lack of institutional capacity and technical knowledge, coupled with donor dependency, unrealistic political commitment, improper use of subsidies and a lack of policy and legal framework are often problems faced in the policy and institutional area. While the reliability of system output and autonomy may be tackled through technological advances, careful considerations should also be made on an institutional level to facilitate the expansion of decentralised electrification programmes, enabling local involvement and economic growth.

Policy implications

- Projects for diversifying renewable energy sources should be implemented alongside with each other rather than in sequence of each other;
- Geographical realities must be taken into account and case-by-case selections of renewable sources may work better than a one-size-fits-all solution. Where possible, hybrid systems that complement each other should be considered: wind turbines to cover for cloudy days where PV input is low, and rain harvesters for areas with heavy seasonal rainfall, for instance;
- For decentralised programmes, to maximise positive social impact a bottom-up approach is preferred with ample local involvement. Foreign knowledge transfers should enable local participation rather than create dependency, and should allow employment opportunities and help foster domestic industry, creating a cascading effect;
- Education and information dissemination is crucial in increasing participation of decentralised systems at the user-end level. The long term economic benefits should be made clear especially when the initial setup cost of the decentralised system may seem off-putting to many households;
- Steps should be taken to lessen the initial economic burden of decentralised systems via government subsidies or microcredit institutions. A mixture of private and public sector efforts may complement each other where coherent planning and distributive mechanisms are lacking;

- Systems should be linked with productive use of the energy to foster local economic growth and flow. It should also support the user's ability to pay for the programme in the long run.

CONCLUSION

This paper has highlighted the difficulties in practically realising decentralised power systems through relying on single renewable power source. Instead, hybrid power systems that are sensitive to local conditions should be considered. The potential employment of rain energy harvesting has been proposed as a complement to current renewable technologies in order to reduce reliance on main electric grid, diesel generators and unclean traditional power sources. The power density potentially achievable from the energy of rainfall is comparable to that of its solar and wind counterparts, and further experimental investigation and case study analysis are required to assess the feasibility of this novel approach. On an implementation front, careful policy and institutional mechanism should be introduced to ensure successful realisation of decentralised power for sustainable development of remote areas.

REFERENCES

1. Wolfe, P., The Implications of an Increasingly Decentralised Energy System, *Energ. Policy*, Vol. 36, No. 12, pp 4509-4513, 2008.

2. Kaundinya, D. P., Balachandra, P. and Ravindranath, H., Grid-connected Versus Stand-alone Energy Systems for Decentralized Power - A Review of Literature, *Renew. Sust. Energ. Rev.*, Vol. 13, No. 8, pp 2041-2050, 2009.

3. Komatsu, S., Kaneko, S. and Ghosh, P. P., Are Micro-benefits Negligible? The Implications of the Rapid Expansion of Solar Home Systems (SHS) in Rural Bangladesh for Sustainable Development, *Energ. Policy*, Vol. 39, No. 7, pp 4022-4031, 2011.

4. van Els, R. H., de Souza Vianna, J. N., and Brasil, A. C. P., The Brazilian Experience of Rural Electrification in the Amazon with Decentralized Generation – the Need to Change the Paradigm From Electrification to Development, *Renew. Sust. Energ. Rev.*, Vol.16, No. 3, pp 1450-1461, 2012.

5. Ruiz, B. J., Rodriguez, V. and Bermann, C., Analysis and perspectives of the government programs to promote the renewable electricity generation in Brazil, *Energy Policy*, Vol. 35, No. 5, pp 2989-2994, 2007.

6. Galdino, M. A. and Lima, J. H. G., PRODEEM - The Brazilian Programme for Rural Electrification Using Photovoltaics, *Proceedings of Rio 02 World Climate & Energy Event*, Rio, Brazil, January 6-11, 2002, pp 77-84.

7. Pereira, E.B., Abreu, S. L., Stuhlmann, R., Rieland, M. and Colle, S., Survey of the Incident Solar Radiation in Brazil by use of Meteosat Satellite Data, *Sol. Energy*, Vol. 57, No. 2, pp 125-132, 1996.

8. De Souza, J. L., Nicacio, R. M. and Moura, M. A. L., Global Solar Radiation Measurements in Maceio, Brazil, *Renew. Energ.*, Vol. 30, No. 8, pp. 1203-1220, 2005.

9. Costa, M.H. and Foley, J.A., A comparison of precipitation datasets for the Amazon Basin, *Geophys. Res. Lett.*, Vol. 25, No. 2, pp 155-158, 1998.

10. Balijepalli, V. S. K. M. and Khaparde, S. A., Smart and Sustainable Energy Systems for Developing Countries: An Indian Perspective, *Proceedings of IEEE Power & Energy Society General Meeting*, San Diego, CA, USA, July 24-29, 2011, pp 1-8.

11. Ghosh, S., Das, T. K., and Jash, T., Sustainability of Decentralized Woodfuel-based Power Plant: an Experience in India, *Energy*, Vol. 29, No. 1, pp 155-166, 2004.

12. Moharil, R. M. and Kulkarni, P. S., A Case Study of Solar Photovoltaic Power System at Sagardeep Island, India, *Renew. Sust. Energ. Rev.*, Vol. 13, No. 3, pp 673-681, 2009.

13. International Energy Agency, *World Energy outlook 2012*, OECD/IEA, Paris, 2012.

14. Sadrul Islam, A. K. M., Islam, M., and Rahman, T., Effective Renewable Energy Activities in Bangladesh, *Renew. Energ.*, Vol. 31, No. 5, pp 677-688, 2006.

15. Shaahid, S. M. and Elhadidy, M. A., Economic Analysis of Hybrid Photovoltaic-diesel--battery Power Systems for Residential Loads in Hot Regions - A Step to Clean Future, *Renew. Sust. Energ. Rev.*, Vol. 12, No. 2, pp 488-503, 2008.

16. IDCOL, *Renewable Energy projects*, URL: http://www.idcol.org/energyProject.php, [Accessed: 6-Sep-2013]

17. Rahman, M. Z., Multitude of Progress and Unmediated Problems of Solar PV in Bangladesh, *Renew. Sust. Energ. Rev.*, Vol. 16, No. 1, pp 466-473, 2012.

18. Islam, R. M., Islam, R. M., and Beg, A. R. M., Renewable Energy Resources and Technologies Practice in Bangladesh, *Renew. Sust. Energ. Rev.*, Vol. 12, No. 2, pp 299-343, 2008.

19. Warner, C. L., Taylor, R. W., Ribeiro, C. M., Moszkowicz, M., and Borba, A. J. V., PV-hybrid Village Power Systems in Amazonia, *Proceedings of the 25th IEEE Photovoltaic Specialists Conference*, Washingtong D.C., May 12-17, pp 1469-1472, 1996.

20. Borges Neto, M. R., Carvalho, P. C. M., Carioca, J. O. B., and Canafístula, F. J. F., Biogas/photovoltaic Hybrid Power System for Decentralized Energy Supp.ly of Rural Areas, *Energ. Policy*, Vol. 38, No. 8, pp 4497-4506, 2010.

21. Schmid, A. L. and Hoffmann, C. A. A., Replacing Diesel by Solar in the Amazon: Short-term Economic Feasibility of PV-diesel Hybrid Systems, *Energ. Policy*, Vol. 32 No. 7, pp 881-898, 2004.

22. Kemmoku, Y., Ishikawa, K., Nakagawa, S., Kawamoto, T., and Sakakibara, T., Life Cycle CO_2 Emissions of a Photovoltaic/Wind/Diesel Generating System, *Trans. Inst. Electr. Eng. Jpn. B.*, Vol. 120-B, No. 7, pp 923-930, 2000.

23. Saheb-Koussa, D., Haddadi, M., and Belhamel, M., Economic and Technical Study of a Hybrid System (wind-photovoltaic-diesel) for Rural Electrification in Algeria, *App.l. Energ.*, Vol. 86, No. 7-8, pp 1024-1030, 2009.

24. Celik, A. N., Optimisation and Techno-economic Analysis of Autonomous Photovoltaic - Wind Hybrid Energy Systems in Comparison to Single Photovoltaic and Wind Systems, *Energ. Convers. Manage.*, Vol. 43, No. 18, pp 2453-2468, 2002.

25. The World Bank, Average precipitation in depth (mm per year), URL: http://data.worldbank.org/indicator/AG.LND.PRCP.MM, [Accessed: 25-Jul-2013]

26. The World Bank, Access to electricity (% of population), URL: http://data.worldbank.org/indicator/EG.ELC.ACCS.ZS, accessed on 10 Feb. 2014.

27. Nandi, S. K., Hoque, M. N., Ghosh, H. R. and Roy, S. K., Potential of Wind and Solar Electricity Generation in Bangladesh, *ISRN Renew. Energ.*, Vol. 2012, Article ID 401761, pp 10, 2012.

28. The World Bank Climate Change Knowledge Portal, Average monthly temperature and rainfall   for   Bangladesh   from   1990-2009,   URL: http://sdwebx.worldbank.org/climateportal/index.cfm?page=country_historical_climate &ThisRegion=Asia&ThisCCode=BGD, [Accessed: 9-Feb-2014]

29. Guigon, R., Chaillout, J.J., Jager, T. and Despesse, G., Harvesting raindrop energy: theory, *Smart Mater. Struct.*, Vol. 17, No. 1, pp 015-038, 2008.

30. Guigon, R., Chaillout, J.J., Jager, T. and Despesse, G., Harvesting raindrop energy: experimental study, *Smart Mater. Struct.*, Vol. 17, No. 1, pp 015-039, 2008.

31. Villemaux, E. and Bossa, B., Single-drop fragmentation determines size distribution of raindrops, *Nautre Phys.*, Vol. 5, No. 9, pp 697-702, 2009.

32. Beard, K. V., Terminal Velocity and Shape of Cloud and Precipitation Drops Aloft, *J. Atmos.   Sci.,   Vol.   33,   pp   851-864,   1976.*

33. Priya, S. and Inman, D. J., *Energy Harvesting Technologies*, Springer: New York, USA, 2009.

34. Lallart, M., Priya, S., Bressers, S. and Inman, D. J., Small-scale Piezoelectric Energy Harvesting Using Low-energy-density Sources, *J. Korean Phys. Soc.*, Vol. 57, No. 4, pp 947-951, 2010.

35. Wang, D. A. and Ko, H. H., Piezoelectric Energy Harvesting from Flow-induced Vibration, *J. Micromech. Microeng.*, Vol. 20, No. 2, pp., 2010.

36. Gao, X., Shih, W. H. and Shih, W. Y., Flow Energy Harvesting Using Piezoelectric Cantilevers with Cylindrical Extension, *IEEE Trans. Ind. Electron.*, Vol. 60, No. 3, pp 1116-1118, 2012.

37. Pozzi, M. and Zhu, M., Plucked Piezoelectric Bimorphs for Knee-joint Energy Harvesting: Modelling and Experimental Validation, *Smart Mat. Struc.*, Vo. 20, No. 5, pp., 2011.

38. Urmee, T., Harries, D., and Schlapfer, A., Issues Related to Rural Electrification Using Renewable Energy in Developing Countries of Asia and Pacific, *Renew. Energ.*, Vol. 34, No. 2, pp 354-357, 2009.

6

Determining the Groundwater Balance and Radius of Influence Using Hydrodynamic Modeling: Case Study of the Groundwater Source Šumice in Serbia

Dušan Polomčić[1], Dragoljub Bajić[*2], Jelena Zarić[3]

[1]Department of Hydrogeology, Faculty of Mining and Geology, University of Belgrade, Đušina 7, Belgrade, Serbia
e-mail: dupol@gmail.com
[2]Department of Hydrogeology, Faculty of Mining and Geology, University of Belgrade, Đušina 7, Belgrade, Serbia
e-mail: osljane@orion.rs
[3]Department of Hydrogeology, Faculty of Mining and Geology, University of Belgrade, Đušina 7, Belgrade, Serbia
e-mail: jelena_zaric@live.com

ABSTRACT

A groundwater flow model was developed to simulate groundwater extraction from the public water supply source of the City of Kikinda. The hydrodynamic model includes the municipal groundwater source of Kikinda (Šumice and the Jezero Well), but also an extended area where there are groundwater sources that provide water supply to three factories: (MSK, TM and LŽT - Kikinda). Hydrodynamic modeling, based on the numerical method of finite differences will show the groundwater balance of the sources in the extended area of Kikinda. The impact of the industrial water sources on the regime of the public water supply source will also be assessed. The radius of influence of the groundwater source is determined by simulating the travel of conservative particles over a period of 200 days.

KEYWORDS

Water supply, Aquifer, Flow, Groundwater regime, Numerical modeling, Particle tracking.

INTRODUCTION

Kikinda Municipality is located in the northeastern part of the Province of Vojvodina and occupies a land area of 782 km^2. According to the most recent census (2002), the population count is 41,935 and the population growth rate -5.7%. Over the past several years, geological and hydrogeological research has been conducted in the extended area of Kikinda's groundwater source, aimed at providing the required amount of quality groundwater to meet the water demand of the population. The study area is situated in the southern part of the Pannonian Basin, which is a lowland that occupies the north of Serbia (Figure 1). The City of Kikinda extends over an area of 13.5 km^2 in the spacious Banat Plain, where the highest elevation is 83 m and the lowest 76 m. Water supply is provided to Kikinda by means of 11 wells. The tapped water-bearing horizon is at a depth of about 250 m, formed in Quaternary sands whose thickness is about 50 m. In the vicinity there are also groundwater sources for industrial water supply, which have a certain influence on the regime of the public water supply source.

[*] Corresponding author

Nowadays, water supply is a highly complex and significant issue in every society. It is of vital importance to control and monitor the operation of each water source. Unfortunately, there are many problems. Even when the required legislation is in place, it is sometimes impossible to implement it in practice, not only in the case of existing sources that have been in service for years, but also when it comes to opening new sources. The best research approach includes hydrodynamic analysis in the early stages of planning and sound monitoring in all stages. Many authors have recently focused on this subject from different perspectives: the hydrodynamic aspect of hydrogeological research [1-3], water source optimization and sustainable development of water supply systems [4-6], drinking water supply [7], different types of factors and processes that influence water source performance [8], water quality [9, 10], and protection of water sources [6, 11].

The groundwater balance of the groundwater sources was determined by hydrodynamic modeling. The impact of industrial groundwater sources on the municipal groundwater source was also assessed. The particle tracking method [12] was used to determine the radius of influence of the groundwater sources over a period of 200 days. The 200-day period reflected the travel time to the wells and represented the third and widest sanitary protection zone [13]. Particle tracking can be used for various purposes. The backward particle tracking method with an uncertainty analysis of porosity, applying a Monte Carlo approach, with Geographic Information System (GIS) support, is a useful tool for delineating groundwater protection zones [14]. Particle tracking simulations can be used to determine travel times from recharge to discharge areas along identified flowpaths [15]. Using a standard numerical flow and transport code and a technique based on adjoint theory, and combining forward-in time and backward-in time transport modeling, it is possible to determine the impact of potential contaminant sources at unknown locations within a well capture zone, including the expected times of arrival of a contaminant, the dispersion-related reduction in concentration, the time taken to breach a certain quality objective and the corresponding exposure times [16].

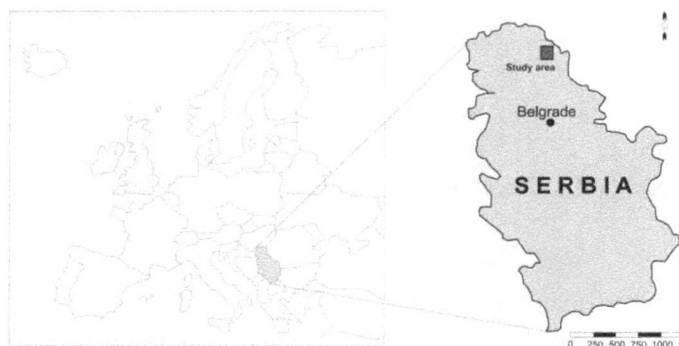

Figure 1. Geographic position of study area

GEOLOGICAL SETTING OF THE STUDY AREA

Extensive geological research (exploratory drilling, geophysical activities, paleontological investigations) has resulted in a relatively robust geological database [17]. The lithological units comprise Pliocene and Quaternary sediments. The base of the Quaternary sediments is made up of shallow lacustrine sediments of the Upper Pliocene. They are built of sandy siltstone, gravelly and sandy siltstone, dark gray siltstone and organogenic silty sands. Carbonaceous interbeds and organic matter are regularly present in these sediments. The color of the sediments ranges from dark gray to light gray and olive gray. Pleistocene and Holocene deposits were abstracted during the Quaternary period. The earliest anthropogenic formations were built from Lower Pleistocene riverine and lacustrine sediments. Riverine/bog sediments of the Mindel period constitute an immediate sequel, whose continuity extended during the Mindel-Riss interglacial period. The Upper Pleistocene is comprised of riverine/bog sediments of the riverbed facies and floodplain facies, which constitute the Varoška Terrace deposits of the Würm age. Sedimentological analyses have shown that the riverine and lacustrine sediments were made of sands, silty sands and alevrites (sandy, rarely clayey and gravelly), while the riverine/bog sediments were composed of clay alevrites and alevrite sands with gravel.

Additional units were abstracted in the post-glacial period, during the Holocene: abandoned channel facies, floodplain facies, bog sediments, alluvia, and the youngest unit deposited since the Holocene – the beach facies of the Tisa River. Sedimentological analyses have shown that they were mostly made up of alevrites, sand, gravel and loess.

HYDROGEOLOGICAL OVERVIEW

A large number of wells (Figure 2) have been drilled in the study area [18]. According to the type of porosity and hydrodynamic characteristics, there are two types of aquifers in this region: confined and unconfined, both in Quaternary sands.

The unconfined aquifer was formed in Quaternary alluvial sands. The depth of the sediments is about 30 m. The aquifer is rechared by infiltration of precipitation and surface water, since there is a good hydraulic contact between the aquifer and the river. The aquifer is drained naturally, at times of low river stages when groundwater flow is directed towards rivers, and artificially - through drilled and dug wells.

The confined aquifer was formed within several water-bearing complexes of Quaternary sands. Although this aquifer comprises a number of water-bearing layers, it was not possible to separate them on the basis of geometry, hydraulic interactions and physico chemical characteristics of the water. The reason for this is the evolution of the terrain, or, more precisely, there are lacustrine, bog and riverine sediments for which there are no reliable data about horizontal and vertical distributions and interbeds of semi-permeable and impermeable strata. An important characteristic of these layers is the constant alternation of sands, slightly gravelly sands, with loessial clays, sandy clays, and clays (Figure 3). Figure 3 shows a characteristic lateral alternation and thinning of impermeable layers and the water-bearing layer. The layers of sand, which are in places more than 30 m thick (although typically less than 10 m), represent the basis of certain polygenic packages. Overlying the sands, there are microgranular sediments: loamy sand, loessial clays, clays, and sandy clays. The transitions to fine-grain sediments are often gradual.

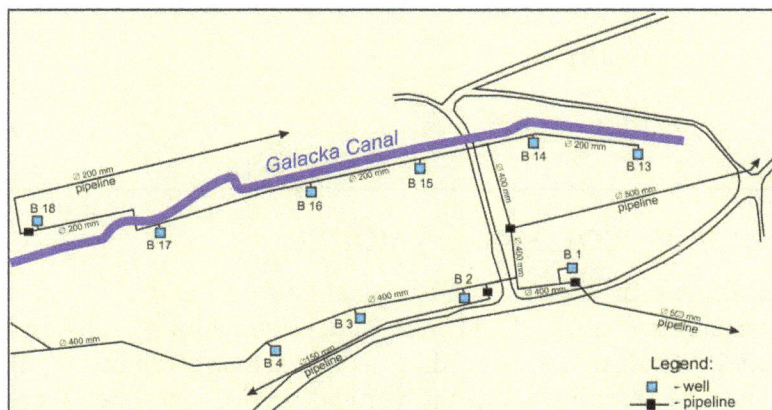

Figure 2. Distribution of wells at the groundwater source of Šumice

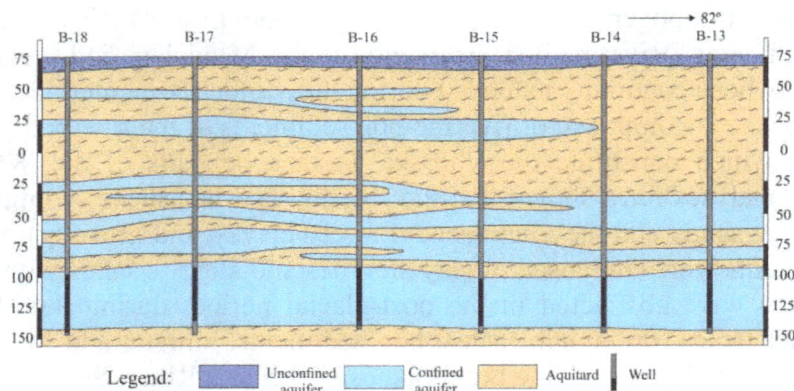

Figure 3. Hydrogeological section through the groundwater source of Šumice

Based on the grain-size distribution of the water-bearing horizons, which also include the studied aquifer, the material of the water-bearing formation is represented by fine- and medium-grain sands (uniformity coefficient: $C_u = d_{60}/d_{10} = $ 2-3 mm with $d_{50} = $ 0.1-0.3 mm, where d is particle/grain size: d_{10}, d_{50} and d_{60} represents a grain diameter for which 10%, 50% and 60% of the sample will be finer than it or using another words, 10%, 50% and 60% of the sample by weight is smaller than diameter d_{10}, d_{50} and d_{60}). According to pumping test data, the transmissibility coefficient in the extended area of Kikinda, depending on the locality in question, is in the range of (2.5-10) $\times 10^{-2}$ m^2/s.

Table 1 shows the hydrogeological parameters previously determined for the Kikinda area [19]. The aquifer extends beyond the study area. The aquifer is probably recharged on the slopes of the Carpathian Mountains, the Vršac Mountains and Mt. Fruška Gora (east of the study area). It is also recharged from surface water in the southwestern part of the area, where the depth of the layers which from the aquifer is relatively small and allows a hydraulic contact between the river and the aquifer over the alluvial layers. The main types of aquifer drainage are: artificial drainage (through groundwater extraction) and leakage into overlying semi-permeable strata (given that the aquifer is confined), and from thereto the top aquifer. In lithological terms, aquitards are generally represented by clays and sandy clays (Figure 3).

Table 1. Hydrogeological parameters from previous research

Well	Transmissibility coefficient [m^2/s]	Hydraulic conductivity [m/s]	Year
Šumice (B-2)	(10-12) $\times 10^{-3}$	-	1979
TM	(11-18) $\times 10^{-3}$	-	1979
MSK	(3-7) $\times 10^{-3}$	-	1978
MSK	(17-22) $\times 10^{-3}$	-	1981
MSK	20.5 $\times 10^{-3}$	8.5 $\times 10^{-3}$	1995

CONCEPTUAL (HYDROGEOLOGICAL) MODEL

The conceptual model of the hydrogeological system is based on geological information on the boreholes (Figure 2) and water-level fluctuations in observation wells. The conceptual model was designed according to the actual groundwater flow in the basin. It was simulated by three layers in the vertical section and one layered aquifer, where the horizontal and vertical flows between the simulated layers were considered. The model layers, from the ground surface down, are shown in Table 2.

Table 2. Model layers and corresponding terrain layers

First confining stratum	Clayey sediments with sand interbeds
Second water-bearing stratum	Fine- and medium-grain sandy sediments
Third confining stratum	Clays and clayey sands

The model reflects the presence of semi-permeable overlying and underlying sediments and a water-bearing horizon, tapped by the previously mentioned groundwater sources. As such, the influence of any hydraulic contact with sandy water-bearing strata that overlie the tapped horizon is avoided, because of both insufficient analysis andartificially-created groundwater renewal difficulties, which certainly affect groundwater source performance. The layer contours were geometrized and transformed into the coordinate system of the model based on voluminous borehole data collected in the extended area. Figure 4 shows the three-dimensional hydrogeological model of the sectional view of Šumice. The tapped water-bearing horizon is shown in blue, with production well screens shown in green. As previously mentioned, the overlying and underlying strata of this water-bearing horizon are comprised of semi-permable and impermeable sediments.

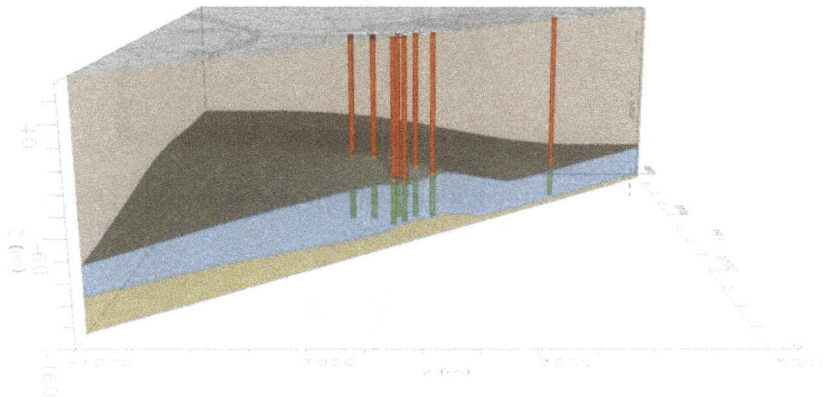

Figure 4. 3D hydrogeological model of the groundwater source of Šumice, 230° azimuth section

HYDRODYNAMIC MODEL

The concept of the hydrodynamic model of the groundwater sources that provide water supply to Kikinda is based on the simulation of three-dimensional groundwater flow. Natural factors, such as the type and characteristics of the represented geological units, the distribution of water-bearing and impermeable units, the seepage characteristics of porous media and the mechanism and regime of groundwater flow, as well as the goal of the task at hand, were of primary importance in selecting the mathematical model concept. A multi-layered model was developed. The hydrodynamic model represented the municipal groundwater source of Kikinda (the source at Šumice and the Jezero Well), but also the extended area including the groundwater sources that service the following factories: MSK (producer of methanol and acetic acid), TM (tile producer), and LŽT - Kikinda (manufacturer of industrial machinery).

The tree-dimensional (3D) finite-difference numerical model for the present study was developed using Modflow [20] with Groundwater Vistas as the graphical user interface [21]. The model encompassed an area of 6 km × 6 km and a depth of 300 m. Its orientation was north-south and the discretization 100 m × 100 m in three layers. Grid cells size was refined to 12.5 m × 12.5 m in the area of interest, wherewell density and extraction rates were high (Figure 5). To illustrate the flow field discretization and the derived schematization of the geometric relationships between the lithological elements in the study area, Figure 6 shows the result of schematization in the vertical section.

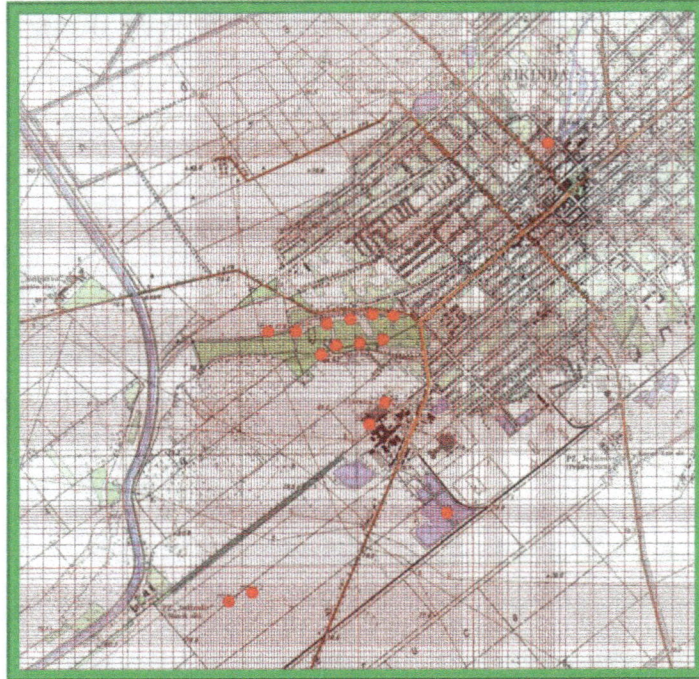

Figure 5. Discretized model and boundary conditions: red – production wells, green contours – general head boundary

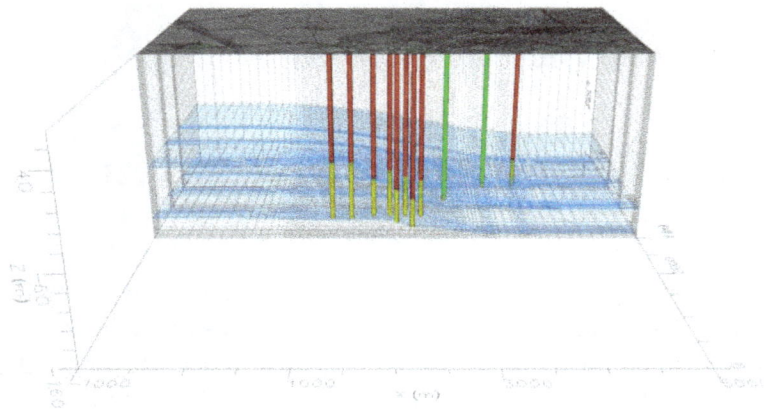

Figure 6. Discretization of the groundwater source at Šumice along the west-east section
(Legend: blue network – aquifer)

Hydraulic parameters

The seepage characteristics of the model layers were specified by the values of hydraulic conductivity, storage coefficient, specific storage, and effective porosity. These hydrogeological parameters were assigned as representative values to each cell. The initial values of hydraulic conductivity of the model were those obtained from earlier tests of the production wells. During the course of the present research, no well pumping tests were conducted for the purposes of the hydrodynamic model, given that they would have affected the municipal and industrial water supply. The initial values of the storage coefficient, specific storage, and effective porosity were adopted based on international reports [22-25], related to the hydrogeological properties of sediments with similar characteristics.

Boundary conditions

The following boundary conditions were specified in the hydrodynamic model of Šumice: head-dependent flux boundary condition (Cauchy or mixed conditions) and boundary of prescribed flux (Neumann conditions).

The head-dependent flux boundary condition (Cauchy or mixed conditions) - the influence of the Galacka Canal, was simulated using this boundary condition. Given the depth of the pumped aquifer at the source and the thick package of semi-permeable and impermeable sediments in its overlying bed, the surface streams have no affect on the groundwater regime of the tapped horizon. This boundary condition was specified in the first layer at locations where both canals exhibited minimum multiannual values, or 74 m.a.s.l. (metres above sea level). The canal widths were specified according to a topographic map, with an elevation 0.5 m less than the minimum canal water level, and the thickness of the bottom canal layer of 0.15 m, with a hydraulic conductivity of 1×10^{-6} m/s.

Using the same boundary condition (head-dependent flux boundary condition), the effect of the source of recharge/point of drainage, located outside of the model area covered, was simulated. In the Modflow code applied here, it was represented by the "general head boundary". This approach was used to set the registered piezometric levels of the pumped water-bearing horizon in the second layer, given the remote locations where this horizon is recharged. This type of boundary condition is shown in Figure 5.

The prescribed flux boundary (Neumann conditions), or the impact of the wells on the groundwater source, was simulated through a specific flux boundary. In this case, the flux was specified as a function of position and time. Figure 5 shows the positions of the wells, whose operation was simulated in the model with a specified prescribed-flux boundary condition. For the purposes of the hydrodynamic model, well discharges at Šumice were registered every seven days during the period from January 1 to December 31 (Table 2). Figure 7 shows individual well discharges recorded at Šumice during the one-year period.

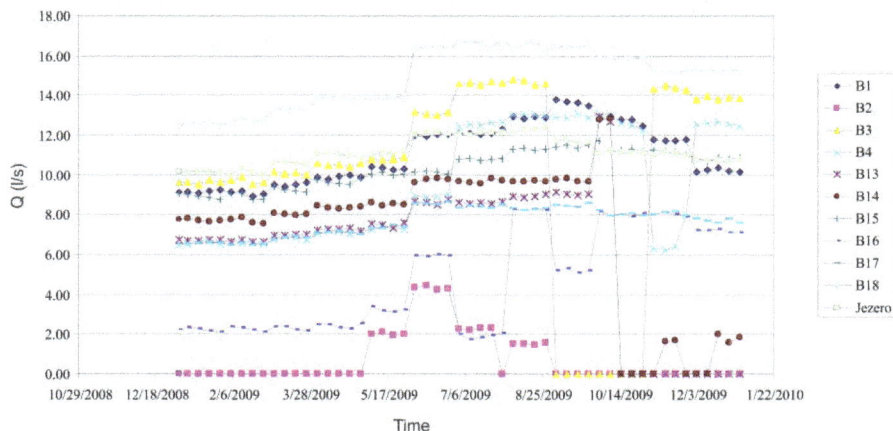

Figure 7. Registered individual well discharges, January 1 to December 31, 2009

The individual well discharges in the model of Šumice were specified in accordance with both the recorded operating hours and discharge rates. The well screens were specified using as-built dimensions and positions. The discharge rates of the production wells of the other groundwater sources in the area were specified based on available technical reports [26], as constant values throughout the period. These included the MSK groundwater source, operating two production wells with a total capacity of 54 l/s; the TM groundwater source, using a single production well with a discharge rate of 35 l/s; and the LŽT - Kikinda groundwater source, which operated two production wells with a total capacity of 50 l/s.

Effective infiltration. In the overall groundwater balance, for the reasons stated in the description of the boundary condition "rivers", the so-called vertical balance had no direct impact on the tapped water-bearing horizon. It was simulated in the first layer by variable-flux Neumann boundary conditions. This value is the sum of infiltrated precipitation and evapotranspiration. The depth-to-groundwater during the entire study period (January 1 to December 31, 2009) was several meters below the ground surface (4-7.3 m). Bearing in mind that the top layer is semi-permeable, the impact of rainfall infiltration on the groundwater regime is small. Ten-percent infiltration was used as the initial value of effective infiltration. Figure 8 shows mean monthly precipitation levels for the period from 1996 to 2006.

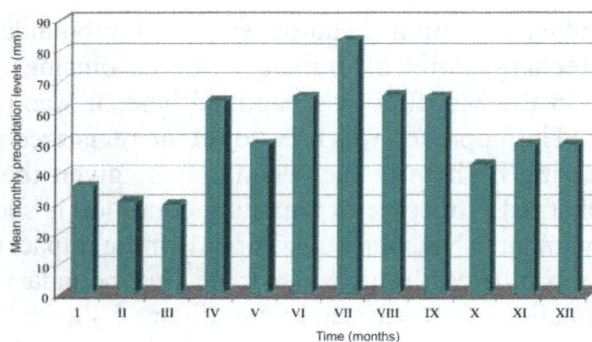

Figure 8. Mean monthly precipitation levels for the period 1996-2006

Model calibration

The model was calibrated under the conditions of transient flow, with a time step of seven days for the study period (January 1 to December 31, 2009). Figure 9 is a parallel representation of the total capacity of Šumice and the registered piezometric levels in the observation wells.

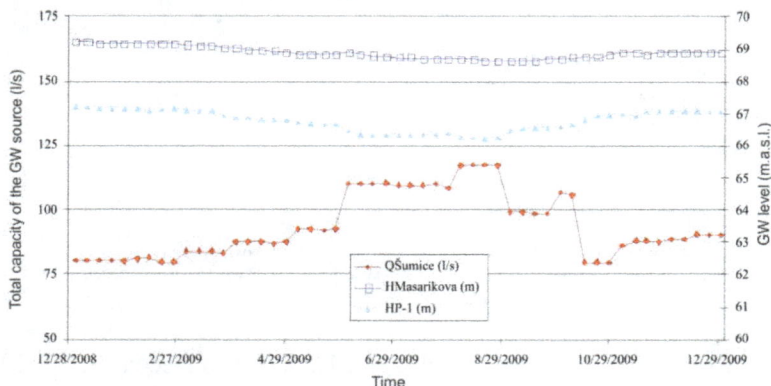

Figure 9. Total capacity of the groundwater source at Šumice and piezometric levels in observation wells (January 1 to December 31, 2009)

The calibration of the model was deemed completed when a satisfactory match between registered groundwater levels and those obtained by calculations was obtained. Figure 10 shows the distribution of piezometric levels in the tapped water-bearing horizon for the maximum rate of groundwater extraction at Šumice of 117.66 l/s (August 20, 2009). Figure 11 shows the groundwater levels registered at the observation wells and those obtained by calculations in the model calibration process (using the same observation wells). The agreement of the registered and calibrated groundwater levels was rather good.

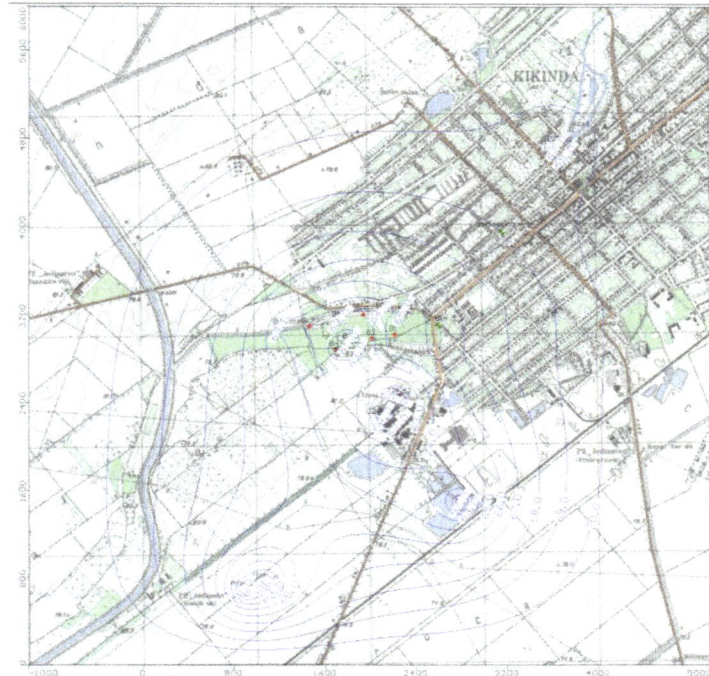

Figure 10. Piezometric head distribution in the extended area of the groundwater source of Šumice (August 20, 2009), at maximum capacity

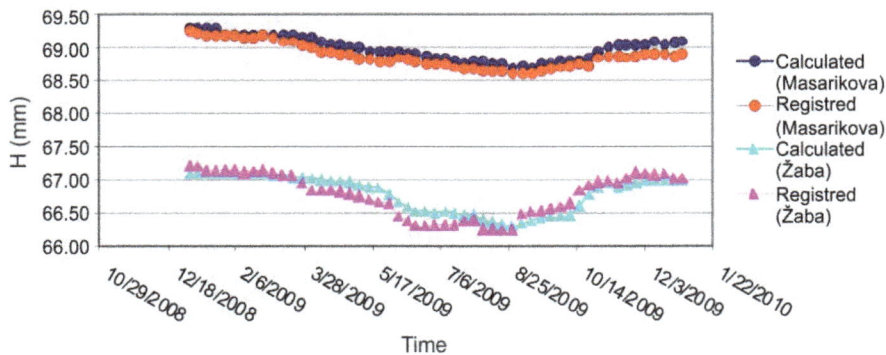

Figure 11. Groundwater levels in piezometer P-1 Žaba and Masarikova Well acting as an observation well, registered and calculated during the model calibration process

GROUNDWATER BALANCE

Assessment of the groundwater balance of the area covered by the model revealed that under the conditions that prevailed during maximum groundwater pumping at Šumice, most of the groundwater in the tapped water-bearing horizon came from the south (35.23%). Table 3 shows the inflow into the area covered by the model under the given conditions (August 20, 2009), or, in other words, the summary groundwater balance of municipal (117.66 l/s) and industrial (139.00 l/s) sources of water supply.

Table 3. Groundwater balance

	Model inflow [l/s]	Model outflow (wells) [l/s]
North	27.57	
East	73.85	256.66
South	90.38	
West	64.72	
Sum	256.52	256.66

RADIUS OF INFLUENCE OF THE GROUNDWATER SOURCE AT ŠUMICE

Modpath [12] was mostly used to simulate the conservative advection of dissolved phase contaminants or microbes within the groundwater system over selected periods, ignoring the effect of dispersion. Hydraulic heads from Modflow [20] were used and pore water velocity computed based on hydraulic conductivity and porosity data. The three-dimensional models easily allowed the consideration of layer heterogeneity and partial penetration of the well screens. As previously stated, particle tracking is suitable for different purposes. In this case, the Modpath code was used to assess the radius of influence of the Šumice wells. The distance from which groundwater reached the wells in 200 days was determined. The streamline arrows pointing to the production wells denote the 20-day travel time of a conservative particle. Figure 12 shows that the radius of influence of Šumice (blue contour) is 700-750 m and that of the Jezero Well 350 m.

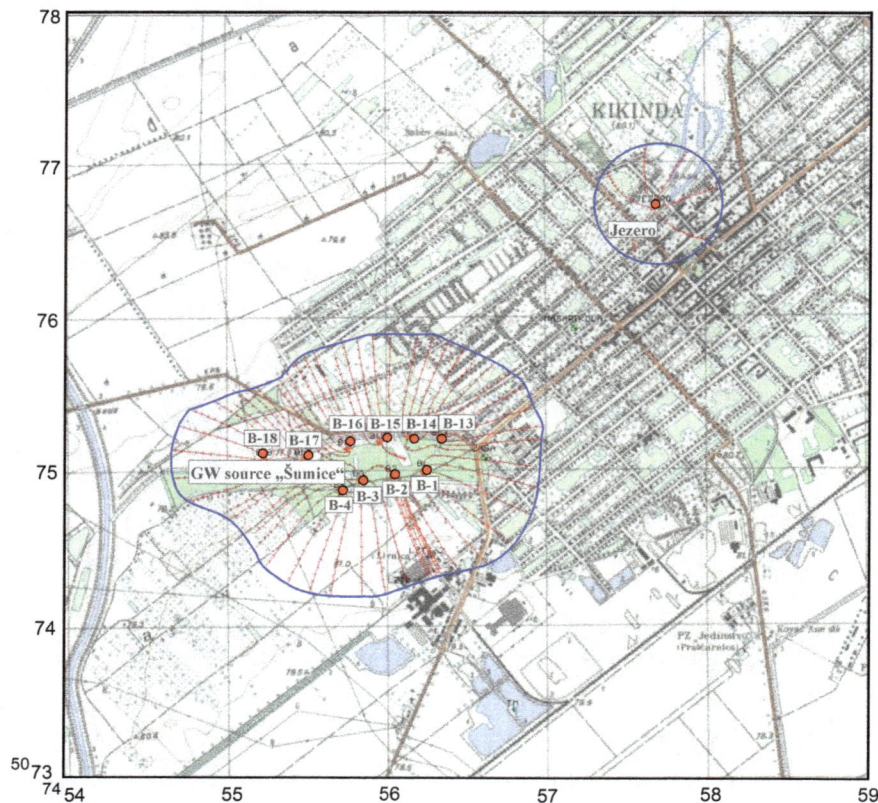

Figure 12. Disribution of streamlines around production wells at the groundwater source of Šumice at a discharge rate of 117.66 l/s, which indicates the distance from the well covered in 200 days

Any paper that does not address new and innovative aspects of the topics of the meeting may be rejected by editors before entering the review process. In order to process the reviewing in time, please submit your manuscript via web interface in camera-ready form and in Microsoft Word format.

CONCLUSION

A groundwater flow model was developed and calibrated for a Quaternary aquifer under unsteady-state conditions. Assessment of the groundwater balance revealed that most of the groundwater flow to the water-bearing horizon of the groundwater source at Šumice came from the south (35.23%). Assuming Darcian unsteady-state three-dimensional flow and based on the major hypotheses of this method (linear form of the sink term in the mass transfer equation and negligible order of magnitude of dispersion effects), the particle tracking method for the determination of the radius of influence of the groundwater source was used and presented. The maximum well discharge capacity of Šumice was found to be 117.66 l/s, at which the radius of influence of Šumice was 700-750 m and that of the Jezero Well 350 m.

ACKNOWLEDGMENTS

Our gratitude goes to the Ministry of Education, Science and Technological Development of the Republic of Serbia for financing projects OI-176022, TR-33039 and III-43004.

REFERENCES

1. Hajdin, B., Polomčić, D., Stevanović, Z., Bajić, D. and Hajdin, K., Asseings Prospect of Groundwater Source „Vić Bare" for Obrenovac's Water Supply, *Proceedings of the XIV Serbian Symposium on Hydrogeology*, Zlatibor, Serbia, 17-20 May, 2012, University of Belgrade - Faculty of Mining and Geology: Belgrade, pp 107-111, 2012.
2. Polomčić, D., Bajić, D., Ristić-Vakanjac, V., Čokorilo, M., Drašković, D., Špadijer, S., Hydrodynamic Characteristics of the Water Supply Source „Peštan" (Lazarevac, Serbia), Vodoprivreda, 261-263, Serbian Irrigation and Drainage Society, Belgrade, pp 55-68, 2013.
3. Đurić, D., Lukić, V., Soro, A., Hydrodynamic assessment of the Expansion of the Groundwater Source of Petrovaradinska Ada in Novi Sad (in Serbian), Vodoprivreda, 258-260, Serbian Irrigation and Drainage Society, Belgrade, pp 265-272, 2012.
4. Polomčić, D., Hajdin, B., Stevanović, Z., Bajić, D., Hajdin, K, Groundwater Management by Riverbank Filtration and an Infiltration Channel: The Case of Obrenovac, Serbia, *Hydrogeology Journal*, Vol. 21, No. 7, pp1519-1530, 2013.
5. Stauder, S., Stevanović, Z., Richter, C. et al., Evaluating Bank Filtration as an alternative to the current Water Supply from Deeper Aquifer: A Case Study from the Pannonian Basin, Serbia, *Water Resource Management*, Vol. 26, No. 2, pp 581-594, 2012.
6. Polomčić, D., Đekić, M., Milosavljević, S., Popović, Z., Milaković, M., Ristić Vakanjac, V., Krunić, O., Sustainable use of Groundwater Resources in Terms of increasing the Capacity of Two Interconnected Groundwater Sources: A Case Study of Bečej (Serbia), *Proceedings of the 11th International Multidisciplinary Scientific Geoconference*, Alabena, Bulgaria, 20-25 June, 2011, STEF92 Technology Ltd., Sofia, Bulgaria, pp 599-606, 2011.
7. Polomčić, D., Stevanović, Z., Dokmanović, P., Ristić-Vakanjac, V., Hajdin, D., Milanović, S., Bajić, D., Optimization of Groundwater Supply in Serbia, *Proceedings of*

the XIV Serbian Symposium on Hydrogeology, Zlatibor, Serbia, 17-20 May, 2012, University of Belgrade, Faculty of Mining and Geology, Belgrade, pp 15-20, 2012.

8. Polomčić, D., Pavlović, V., Bajić, D., Šubaranović, T, Multiannual effects of Peštan Source Operation in the Function of Pre-dewatering the Future Kolubara Basin Opencast Mines, *Proceedings of the VI International Conference "Coal 2013"*, Zlatibor, Serbia, 02-05 October, 2013, Yugoslav Opencast Mining Committee, Belgrade, pp 259-266, 2013.

9. Mastilović Čoporda, T., Radojević, U., Vakanjac, B., Sorajić, S., Groundwater Quality analysis of the Deep Aquifer in Kikinda, Serbia, *Proceedings of the 11th International Multidisciplinary Scientific Geoconference*, Alabena, Bulgaria, 20-25 June, 2011, STEF92 Technology Ltd., Sofia, Bulgaria, pp 489-496, 2011.

10. Polomčić, D., Krunić, O., Ristić-Vakanjac, V., Hydrogeological and Hydrodynamic Characteristics of Groundwater Sources for the Public Water Supply of Bečej (northern Serbia), *Geološki anali balkanskog poluostvra*, Vol. 72, pp 143-157, 2011.

11. Dimkić, M., Pušić, M., Vidović, D., Đurić, D., Boreli-Zdravković, Đ., Pollution Transport analysis in defining the Sanitary Protection Zones of Groundwater Sources in Alluvial Areas, Vodoprivreda, 264-266, Serbian Irrigation and Drainage Society, Belgrade, pp 203-218, 2013.

12. Pollock, D. W., User's Guide for MODPATH/MODPATHPLOT, Version 3: A Particle Tracking Post-processing Package for MODFLOW, the US Geological Survey Finite-difference Ground-water Flow Model, US Geol Surv Open-File Rep 94-464, 1994.

13. Rulebook about the Method of Determining and Maintaining Sanitary Protection Zones for Water Supply, Official Gazette of Republic of Serbia, No. 92, 2008.

14. Moutsopoulos, N. K., Gemitzi, A., Tsihrintzis, A. V., Delineation of Groundwater Protection Zones by the Backward Particle Tracking Method: Theoretical Background and GIS-based Stochastic Analysis, *Environmental Geology*, Vol. 54, pp 1081-1090, 2008.

15. Yidana, M. S., Groundwater Flow Modeling and Particle Tracking for Chemical Transport in the Southern Voltaian Aquifers, *Environmental Earth Science*, Vol. 63, pp 709-721, 2011.

16. Frind, E. O., Molson, J. W., Rudolph, D. L., Well Vulnerability: A Quantitative Approach for Source Water Protection, *Ground Water*, Vol. 44, No. 5, pp. 732-742, 2006.

17. Soro, A., Dimkić, M., Josipović, J., Hydrogeologic Investigations related to Watersupply in Vojvodina, 100 Years of Hydrogeology in Yugoslavia, University of Belgrade – Faculty of Mining and Geology, pp 101-111, 1997.

18. Matić, I., Project of the Detailed Hydrogeological analysis of the Groundwater in the Municipality of Kikinda Territory, Faculty of Mining and Geology, Belgrade, 2007.

19. Rakić, D., Study on the Groundwater Reserves at the Groundwater Source of "MSK" Factory in Kikinda, Hidrozavod DTD, Novi Sad, 2006.

20. Harbaugh, A. W., Banta, E. R., Hill, M. C. and McDonald, M. G., MODFLOW-2000: The U.S. Geological Survey Modular Ground-Water Model, User Guide to Modularization Concepts and the Ground-Water Flow Process, U.S. Geological Survey Open-File Report 00-92, Reston, VA, USA, pp 121, 2000.

21. Rumbaugh, J. O. & Rumbaugh, D. B., Guide to using Groundwater Vistas: version 5. New York: Environmental Simulations, 2007.

22. Akindunni, F. F. and Gillham, R. W., Unsaturated and Saturated Flow in response to pumping of an Unconfined Aquifer: Numerical investigation of Delayed Drainage,

Ground   Water,   Vol.   30,   No.   6,   873-884,   1992.

23.Anderson, M. P. and Woessner, W. W., Applied Groundwater Modeling: Simulation of Flow and Advective Transport, Academic Press, San Diego, CA, 381p, 1992.

24.Rehm, B. W., Groenewold G. H. and Morin, K. A., Hydraulic Properties of Coal and Related Materials, Northern Great Plains, *Ground Water*, Vol. 18, No. 6, 551-561, 1980.

25.Thierrin, J., Davis G. B., and Baber C., A Ground-water Tracer Test with Deuterated Compounds for Monitoring in situ Biodegresation and Retardation of Aromatic Hydrocarbons, *Ground Water*, Vol. 33, No. 3, 469-475, 1995.

26.Babac, D., Study on the Analysis of the Local Water Resources of the Deep Aquifer of the "Šumice" Water Source in Kikinda, the Aspect of Quality and Quantity and the Definition of Exploitation Conditions for the period of next 20-30 years, Balby International, Belgrade, 2006.

7

Research and Development Financing in the Renewable Energy Industry in Brazil

Muriel de Oliveira Gavira[1,2]

[1]School of Applied Sciences, University of Campinas, Brazil
[2]DepartmentFaculty of Sciences, University of Lisbon, Portugal
e-mail: murielgavira@gmail.com

ABSTRACT

In the last decades, the Brazilian government has put many public policies in place in order to create a favourable environment to promote energy efficiency and clean energy. In this paper we discuss the use of research and development financing support by the clean energy industry in Brazil. To do so, we carried out an empirical research analysing secondary data from legislation, literature case studies, and public and industry reports in order to determine if the companies of the clean energy industry have public financial support to research and development. Our ongoing research shows that, despite incentives to stimulate the dissemination of clean energy, the participation of some of the clean energy is very small (especially solar). We believe that the contributions of this study will assist policy makers, and the whole industry, to improve clean energy research and development investments in Brazil.

KEYWORDS

Clean energy, Public policy, Energy policy, Innovation, Sustainable development.

INTRODUCTION

The energy industry has several challenges to face in the next years. The demand for energy is heavily growing in the last years, as a consequence, its exploration, production and distribution costs. Other important challenges are energy security and availability, energy dependency, sustainable development, social justice, etc.

Governments frequently put in place public policies to help society to face such important issues [1]. One way to face many of those challenges is investing in energy efficiency and in clean and renewable energy. These policies deal, mainly, with the supply side of the problem and are able to give diversity of sources reducing energy foreign dependency, energy inequality, harmful environment impacts, and increasing social development [2-5].

Between the instruments of energy policy there are: research and development (R&D), financing business, tax incentives, voluntary agreements, information dissemination, market reforms, pricing and taxation, consumer awareness, human resources education and training, standard and regulations, and others.

In the last decades, the Brazilian government has put many public policies in place in order to create a favourable environment and to promote energy efficiency and clean energy. Currently, the public sector in Brazil has been directing the resources to the creation of a favourable environment for energy efficiency and renewable energy initiatives.

The biggest and most successful Brazilian policy regarding the renewable energy industry was the PROALCOOL (Alcohol National Program). As a result, Brazil is today one of the leaders in the use of biomass to generate energy (electricity and transportation).

Apart from the sugarcane industry, the government has put in place policies and incentives to clean energy sources such as solar, wind, and biofuel. As Brazilian government also intends to promote the technological development of the country, these incentives include financial support to research and development (R&D), such as the Program of Incentives for Alternative Electricity Sources (PROINFA) and the Technological Development Program for Biodiesel.

In the case of R&D policy, governments can support those activities through financing incentives, governmental purchase of high technological goods and services, public research, rights of intellectual propriety, and human resources for innovation.

To meet the energy challenges we need to develop new technologies and to innovate in ways of efficiently use of the present ones. Therefore, investing in R&D for new and improved clean energy systems is fundamental to the adoption of clean energy technologies.

In the energy industry, is particularly relevant to invest in energy efficiency, computing, information technologies, grid management, and low-carbon technologies and process. Also very important is to expand financing instruments to basic science in the areas of fuel cells, hydrogen, advanced renewable energy, modern biofuels and energy storage [6].

In this paper we discuss the results of an ongoing research about R&D financing to renewable energy projects in Brazil and to do so, we carried out an empirical research analysing secondary data from legislation, literature case studies, and public and industry reports in order to determine if the companies of the renewable energy industry, especially new renewable power generation, have benefit from governmental financial support to R&D projects.

RENEWABLE ENERGY IN BRAZIL

It is important to know the main indicators of the renewable energy industry in Brazil in order to understand where are the technological and productive gaps in the sector.

According to IBGE [7] 191 million people lived in Brazil in 2010 and 98% of them had access to the power grid (2008 data). With a population of this size and the exhaustion of the hydropower potential, it is crucial to seek for new sources of energy in Brazil and to promote energy efficiency and distributed forms of electricity generation.

Brazil relies heavily on clean energy sources: about 46% of the country's energy comes from renewable sources, and with the main source of power energy is being hydroelectric power. Additionally, as stated before, because Brazil invested early in ethanol as a result of government incentives put in place in the mid-1970s, today, it is a world leader in ethanol exports and in the use of biomass to produce electricity in the industrial sector [8].

According to EPE [4] clean energy sources (wind, biomass, small hydro, etc.) will increase their participation in the electricity sector in Brazil from 47.5% in 2010 to 46.3% in 2020. In 2010, 19.3% of the primary energy production came from sugarcane products, 13.7% from hydropower, 10.2% from wood, and 4.3% from other renewable sources.

Despite the increase of clean energy generation in Brazil, especially biomass and wind (Figures 1 and 2, and Table 1), the proportion of renewable sources in the total electricity installed capacity has been falling in the last 12 years. Comparing to other countries, Brazil still have a large share of renewable sources (including hydropower) of electricity; however countries such as Germany, France and the United States have increased the participation of renewable sources in their grid.

In addition, in Figure 2 one can see that the participation of new renewable sources, excluding hydropower, and their installed capacity is still very small considering the country's potential. Moreover, fossil fuel thermal energy has been growing faster than the renewable.

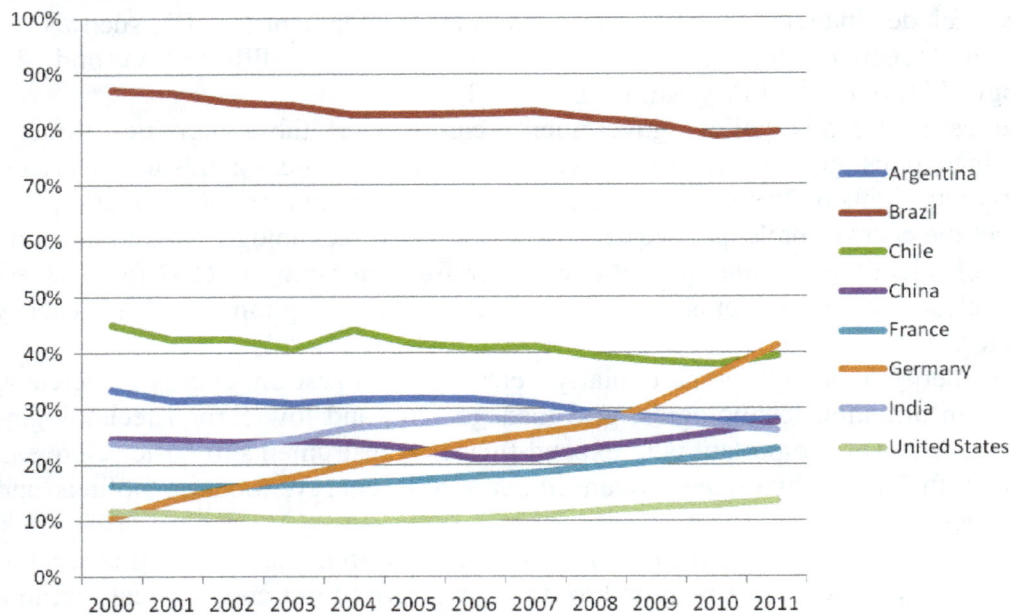

Figure 1. Participation of the renewable sources in the total Electricity Installed Capacity in selected countries
(Source: Elaborated with data from [9])

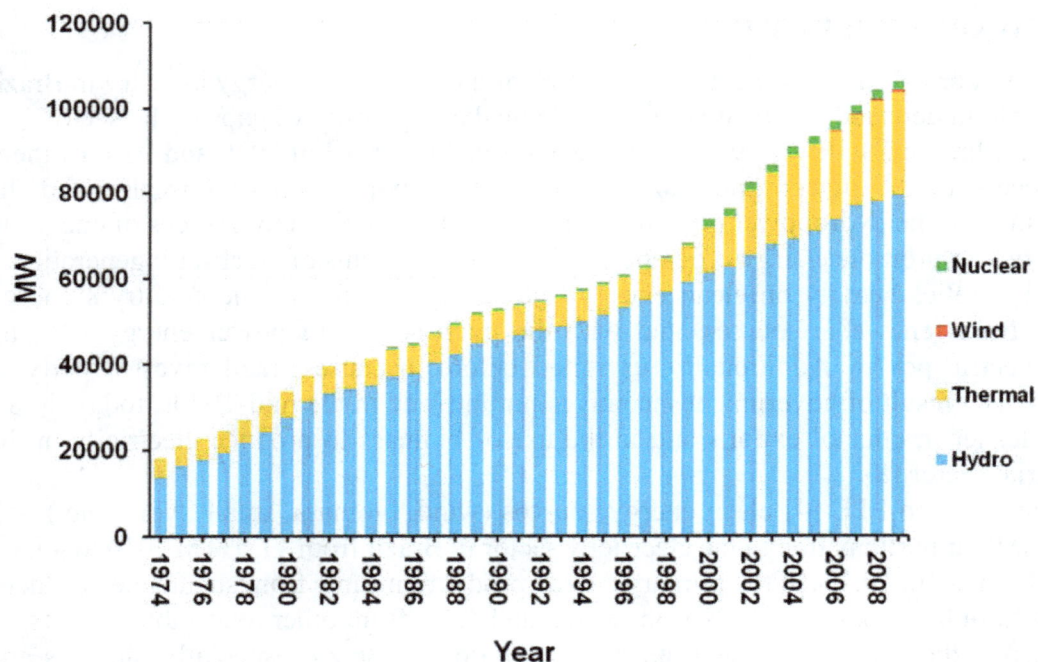

Figure 2. Installed Capacity of Electric Generation [MW]
(Source: Adapted from [4])

On the other hand, Table 1 shows a significant participation of thermal sources in the installed capacity of the country. The Brazilian government have been investing in thermal power plans and their production increase from $2,178 \times 10^3$ tep (toe) in 2000 to $6,845 \times 10^3$ tep (toe) in 2012 [4].

Wind generation has increasing participation in the last years (from 1 GWh in 2000 to 2.176 GWh in 2010) but the wind farms are generally in the Northeast, far from the main consumer centres, due to the climate characteristics and distance from large urban areas.

Table 1. Installed Capacity of Electric Generation [MW] in 2010

Region/Sources	HYDRO			THERMAL			WIND		
	SP	APE	Total	SP	APE	Total	SP	APE	Total
Brazil	77,318	3,385	80,703	17,548	12,141	29,689	926	2	928
North	10,866	29	10,895	3,029	365	3,394			
Northeast	10,776	167	10,943	3,967	1,953	5,920	722	2	724
Southeast	22,661	1,892	24,553	6,034	7,662	13,695	29		29
- São Paulo	10,442	542	10,984	1,145	4,714	5,859			
South	22,042	1,143	23,186	3,178	1,006	4,185	175		175
Centre-West	10,972	154	11,126	1,340	1,156	2,496			

Notes: SP - Public Service. APE - Self-Producers (excluding the partnership between hydroelectric plants with Public Service concessionaries as: Igarapava, Canoas I and II, Funil, Porto Estrela, Machadinho and others).

Source: Elaborated with data from [4]

Other characteristics of the Brazilian Electricity System are [5, 10, 11]:

- The distribution of power is uneven: the access is smaller more limited in the North and Northeast regions; but also these regions have a much smaller population;
- The privatization in the sector started in the mid 1990's;
- The transmission network is integrated (National Integrated System in the states of South, Southeast and Northeast of Brazil). This integrated system cover about 98% of the electricity demand of Brazil and is dominated by hydropower plants with large reservoirs. There are also isolated systems in the North of Brazil, mostly thermal;
- In 2010 there were 98,648.32 km of transmission lines in the National Integrated System (an increase of 3% compared to 2009);
- Power generation and transmission are, for the most part, State controlled. Around 60% of the power distribution (installed capacity) is managed by private companies. Figure 3 shows that the generation is dominated by public utility power plants.

In the world, Brazil is number four in renewable power capacity when hydropower is included [11], and number five in terms of annual capacity increment. Such an increment is largely due to ethanol and biodiesel products (Brazil is number two in biomass power), with wind and solar PV power still being incipient. Brazil is developing new projects related to wind power, solar power, and tidal and wave energy, in order to explore its largely untapped potential [4].

The expansion of the power generation from cleaner sources is promising for Brazil, especially considering that the country can benefit from carbon trading with countries with emission goals. The case of ethanol, which is already exported to several countries, is particularly relevant. But, to tap such a potential, the country needs to develop more sustainable ways of growing sugar-cane and producing ethanol.

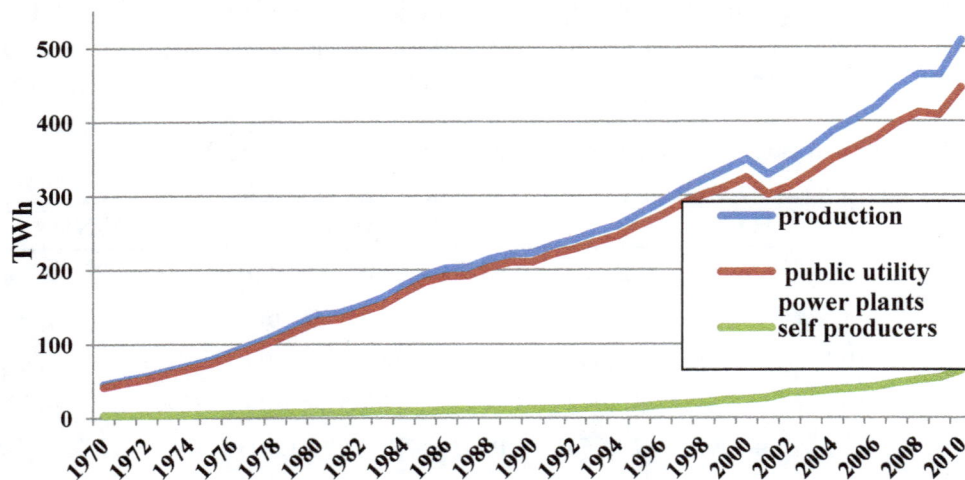

Figure 3. Electricity production in Brazil (flow)
(Source: Elaborated with data from [4])

POLICY INSTRUMENTS FOSTERING RENEWABLE ENERGY IN BRAZIL

Market and technological uncertainties to which clean energy investments are subject may be mitigated by a good institutional and regulatory environment. Well-planned policies and support mechanisms help actors to manage risks and opportunities in the energy industry. In this context, R&D has an important role in promoting renewable electricity since it might result in new and improved innovation to the industry.

Enzensberger et al. [10] divide the policy instruments fostering renewable energy is two larger groups: legislative measures and non-legislative measures. We adapted Enzensberger et al.'s typology to consider R&D financing (Figure 4) and list some examples of instruments:

- Demand and control;
  - o R&D: mandatory investments in R&D (utilities in Brazil must invest 0.5% of their Net revenue in R&D – Lei nº 9.991/2000);
  - o Other instruments: mandatory investments in general, shut-downs, mandatory fuel off-take.
- Construction incentives;
  - o R&D Financing: low or no interested financing;
  - o Other instruments: accelerated depreciation, subsidies, tax deduction and low interested loans.
- Production incentives: fixed feed-in tariffs, tax exemption, actions, etc.;
- Demand-pull: renewable portfolio standard with certificate trading, tax deductions for purchase; public purchase of technology and energy;
- Voluntary: certification, self-goal, etc.;
- Informative or administrative: improvement of administrative process; investor advising; publicity; resource mapping;
- R&D: investments in R&D for energy efficiency and renewable energy.

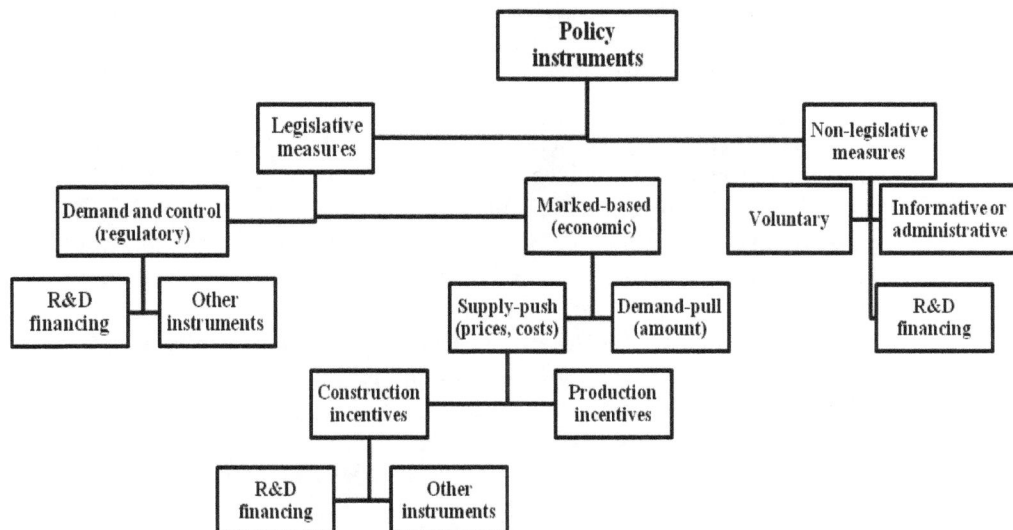

Figure 4. Typology of policy instruments
(Source: adapted from [10])

Research and development (R&D) policies aim to reduce the cost and risk of those activities in order to improve the quantity and quality of the innovation efforts in the country. Regarding public support, the government might offer support instruments such as:

- Tax incentives such as discounts on income taxes based on the amount of R&D investments;
- Taxation over carbon intensive energy sources;
- Direct public financing of R&D projects;
- Financing R&D public institutes that might involve companies in their projects;
- Public purchase of technology and electricity;
- Others: venture capital, public-private partnerships, etc.

In this paper we focus only in direct public R&D financing from market-based (see Figure 4) to the renewable energy industry in Brazil. We do not intend to analyse the R&D investments of the utilities companies in Brazil.

Moreover, the financial incentives to R&D can be classified according to its topic, that is, resources (input), such as labour, machinery and equipment, infrastructure (buildings, information networks, etc.), overhead costs, technological services, materials, etc. Also, it includes incentives that support project development, and output goods and knowledge (licensing process, knowledge management, etc.) [13]. In this paper we focus in all these topics.

Direct public research and development financing

Only in the 1990's Brazil put in place better and reliable instruments of public R&D financing for the private sector. Since then one could see the expansion and diversification of the funding in order to ensure a better resources allocation. However, only part of this goal was, in fact, achieved [11, 13] and Bastos [1]. In Table 2 we present the main sources of public financing to the energy industry in Brazil.

Table 2. Main sources of public R&D financing to the renewable energy industry in Brazil

Type	Institutions
National funding	
Federal Funding agencies	Science and technology agencies from the Ministry of Science, Technology and Innovation (MCTI) such as National Innovation Agency (FINEP); National Council for Scientific and Technological Development (CNPq). Regional development agencies such as Amazon Development Agency (Sudam) and The Superintendence for the Development of the Northeast (Sudene). Regulatory agencies: The Brazilian Electricity Regulatory Agency (Aneel); National Agency of Petroleum, Natural Gas and Biofuels (ANP); etc.
Federal program	Technological support program for exports (Progex), Constitutional Fund for Financing the Centre West Region; etc.
State programs	State funding to support research such as Fundes (RJ), Fomento (GO), and other research funds.
Research Foundations	Banco do Brasil Foundation (FBB); states foundations such as São Paulo Research Foundation (Fapesp); Research and Development Foundation (FADESP); Foundation to Support Research and Extension (Fapex); etc.
Banks	Banco do Brasil; Brazilian Social and Economic Development Bank (BNDES); etc.
International funding	
Programs	Global Environment Facility, United Nations Environment Programme (UNEP).
Banks	The Inter-American Development Bank (IADB or BID), World Bank, The International Bank for Reconstruction and Development (IBRD).

Source: Adapted from [16, 17]

In Table 3 we present the main support to the clean energy industry in Brazil. One can note a variety of different institutions managing the support programs with similar objectives but different types of support instruments. In this way, it is important to have a central institution to guarantee that those instruments work together for the development and diffusion of relevant innovations in the clean energy industry.

In this paper we focused the programs of one its institution: The National Innovation Agency (FINEP).

Finep is directly responsible for the financing of R&D project in Brazil and it is the main source of R&D financing in the last years [18, 7]. This institution issues loans and grants to public and private institutions in Brazil. The grants come from The National Fund for Scientific and Technological Development (FNDCT). It is also is responsible for managing the most part of the Sector Funds of Science, Technology and Innovation and for programs such as [2, 8]:

- Long-term credit with reduced interest rates;
- Concession of economic subsidy for the R&D activity;
- Fiscal incentives: information law, innovation law, etc.;
- Non-reimbursed resources;
- Seed money;
- Scholarships;
- Programs to support the interaction among universities, public research institutions and companies;
- Develop the venture capital segment in Brazil (project "inovar").

Table 3. Main support programs to the clean/renewable energy industry in Brazil

Program	Management Institution	Type of support	Implanted	Specific to the energy industry	Main objective
PROINFA	Eletrobras	Guaranteed contracts of energy purchase	2002	Yes, biomass, wind and small hydro	Increase the share of renewable energy in the National grid from independent power producers
CT-ENERG	National Innovation Agency (FINEP)	R&D and innovation projects financing	2001	Renewable power generation and Efficient energy use	Energy security and diversification, lower costs, increase quality of services, and increase the competitiveness of domestic technology.
InovaEnergia	Brazilian Social and Economic Development Bank (BNDES)	R&D financing: several instruments such as long- term credit with reduced interest rates; non-reimbursed resources.	2013	Yes, to equipment suppliers to the renewable energy industry.	Support Brazilian companies in the global technological develop-ment and production of the photovoltaic, thermo solar and wind power technologies.
Law 9991/2000	The Brazilian Electricity Regulatory Agency (Aneel)	Mandatory investments in renewable energy and energy efficiency	2000	Yes, renewable energy and energy efficiency	To promote constant innovation to overcome the technological challenges of the power industry.

Finep's objective is to support the creation and development of:
- Technology-based businesses that emerge from research centres and universities;
- High-tech spin-offs from large businesses;
- Technology parks;
- Technology-based business incubators;
- Of innovative Clusters;
- Of private research Centres.

In Figure 5 one can see a significant increase in Finep's disbursements in the last years.

The Brazilian science and technology sector funds aim to expand and ensure the constancy of the R&D financing in Brazil and were created to complement and incentive the R&D development in strategic sectors to the county, such as energy, oil, Amazon, agribusiness, biotechnology, etc.

To this research, the most important funds are: Energy, Infrastructure, and "Verde Amarelo".

To participate the companies wait for the calls for projects and them present the proposals, then Finep evaluate the proposals. The call pays expenses such as R&D infrastructure, services, material, equipment, scholarships, etc.

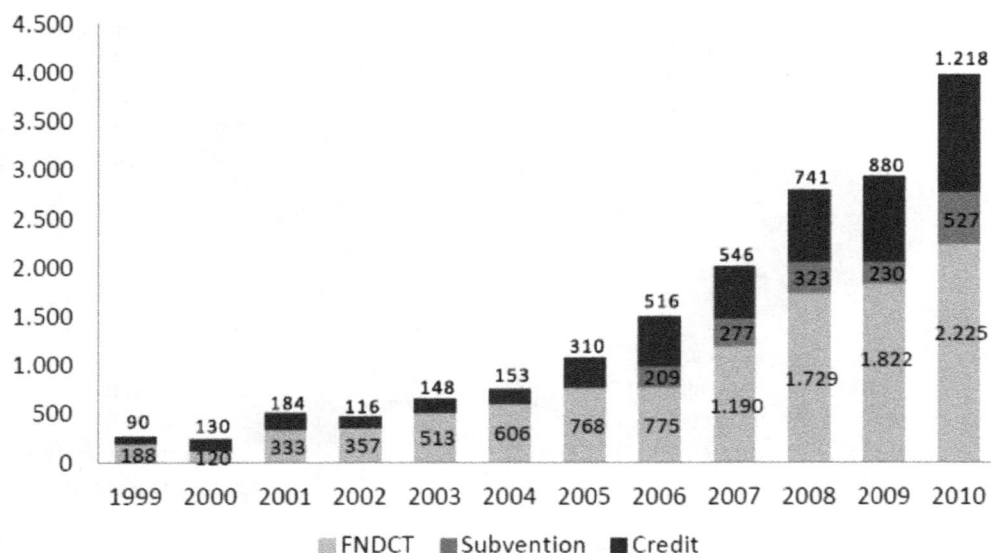

Figure 5. FINEP's disbursements, by program (in local currency - BRL millions), 1999-2010 [19]

The funds Verde Amarelo (FVA) and Infrastructure (CT-Infra) are cross sector and the Energy one (CT-Energ) is specific to the electricity industry.

In recent years Brazil has put in place important initiatives to improve funding opportunities for R&D projects. Major challenges still are the integration between the public funding and develop research and the private sectors and the stability and constancy of funding programs.

The Brazilian financial supports have approximated to those found in other countries, especially OECD countries [2]. These supports have been intensified not only in financing, institutional and incentives resources; but also, in the integration between industrial and economic policies. Those actions aim to increase and improve the integration between the instruments and actors, the knowledge diffusion, the innovation culture, and the competitive advantage derived from technological innovation.

Brazil is far of its goals, once there is still a centralization of recourses and decision making in few institutions, especially FINEP (Financiadora de Estudos e Projetos - Research and Projects Financing) and BNDES (Banco Nacional de Desenvolvimento Econômico e Social - Brazilian Development Bank).

Another issue of considerable importance is the consistency and stability of funding and incentives. That has not happened in recent years because of the resources subordination to national tax problems [2, 19].

CONCLUSION

Well-planned policies and institutional mechanisms can help actors to address risks and exploit opportunities. Indeed, market uncertainties and technological risks in sustainable energy investments are very high, giving to the regulatory environment an essential role in support of multilevel initiatives.

Moreover, since the technological costs and risks in the energy industry are very high, the government should support and incentive the sector to develop itself in a more competitive speed than the present one.

In the last years, the Brazilian government has invested in the creation of policies and incentives aimed at energy efficiency and at other clean energy sources such as solar, wind, and biofuel. Countries have also put into place Science and Technology (S&T) Policies to support Education and Research and Development activities. In Brazil, several initiatives are in place, such as the Program of Incentives for Alternative Electricity Sources (PROINFA) and the Technological Development Program for Biodiesel and the Program for the Hydrogen Economy.

Considering public financial support to innovate, in recent years Brazil has taken important initiatives to improve funding opportunities for projects to research, development and innovation. The main challenge lies on the integration between private (firms) and knowledge generation sectors (universities etc.) and on the constancy financing instruments.

Another issue of considerable importance is the consistency and stability of funding and incentives. That has not happened in recent years because of the resources subordination to national budget problems.

Our ongoing research shows that, despite incentives to stimulate the dissemination of clean energy, the participation of some of the clean energy is very small (especially solar). In addition, the financial resources that the energy industry use to research and develop new clean technologies is still scarce, especially considering the potential of Brazil. However, that amount has increased in the last years, with focus on energy efficiency and hybrid vehicles the country needs to make a better effort in placing mechanisms for awareness, support (especially financial) and promotion of R&D in the renewable industry.

It is still not clear if those policies bring benefits to the country, such as energy security, diversification of the power grid, reduction of foreign oil dependency, and technological development. It is evident that for the clean energy industry to develop, the country needs to invest more in innovation and in mechanisms of awareness, and support.

The fact that the Brazilian energy grid is considered one of the cleanest in the world may cause the government to accommodate and lose sight of a bigger picture that embeds great opportunities for institutional and technological improvements towards a sustainable energy industry.

We believe that the contributions of this study will assist policy makers, and the whole sector, to improve clean energy research and development investments in Brazil.

The next step of our research is to study the results of research and development founding in Brazil from Finep and BNDES especially from the Sectors Funds and the Program of Incentives for Alternative Electricity Sources (PROINFA).

REFERENCES

1. Bastos, V. D., Public funds for science and technology (in Portuguese), *Revista do BNDES*, Vol. 10, no. 20, pp. 229-260, 2003.

2. Corder, S., Financing and incentives to the innovation system in Brazil: current situation and perspectives, *Ph.D thesis*, UNICAMP, Instituto de Geociências, Departamento de Política Científica e Tecnológica, Campinas, ago. 2004. Retrieved from: http://www.bibliotecadigital.unicamp.br/document/?code=vtls000349489, [Accessed: 12-Dec-09], (in Portuguese)

3. Economist Intelligence Unit - EIU, *Scattering the seeds of invention: the globalisation of research and development*, 2004. Retrieved from:

http://graphics.eiu.com/files/ad_pdfs/RnD_GLOBILISATION_WHITEPAPER.pdf , [Accessed:04-April-12]

4. Energy Research Company - EPE (Brasil), *Brazilian Energy Balance Year 2012*, Rio de Janeiro, 2013.

5. International Energy Agency – IEA, *Renewables in global energy supply: an IEA fact sheet*, International Energy Agency, 2007.

6. Energy Research Company - EPE (Brasil), *National Energy Plan 2020*, Rio de Janeiro, 2011.

7. The Brazilian Institute of Geography and Statistics - IBGE, *2010 Population Census*, IBGE, 2011.

8. Alvarenga, G. V., Pianto, D. M. and Araújo, B. C., Impacts of the Brazilian science and technology sector funds on industrial firms' R&D inputs and outputs: new perspectives using a dose-response function. In: *Proceedings, EncontroNacional de Economia – ANPEC*, Porto de Galinhas (PE): Anpec, 11-14 Dec, 20 12. Retrieved fromhttp://www.anpec.org.br/encontro/2012/inscricao/files_I/i8-5fe1cb9e5d777ea4 0cd1a965ecfba0b8.pdf, [Accessed: 13-Sep-13]

9. U.S. Energy Information Administration – EIA. International Energy Statistics. Retrieved from: http://www.eia.gov/countries/, [Accessed: 28-Feb-14]

10. Enzensberger, N., Wietschel, M., and Rentz, O., Policy instruments fostering wind energy projects-a multi-perspective evaluation approach, *Energy Policy*, Vol. 30, no. 9, pp. 793-801, 2002.

11. REN21, *Renewables 2011 Global Status Report*, Paris, REN21 Secretariat, 2011. Retrieved from: http://www.map.ren21.net/GSR/GSR2012.pdf [Accessed: 06-Nov-12]

12. Heugens, P. P. M. A. R. and Lander, M., Structure! Agency! (and other quarrels): meta-analyzing institutional theories of organization, *Academy of Management Journal*, Vol. 52, No. 1, pp 61-85, 2009.

13. Lundvall, B-Å., *National Innovation Systems: towards a theory of innovation and interactive learning*, Pinter, London, 1992.

14. Nelson, R., *National Innovation Systems: a comparative analysis*, Oxford University Press, New York/Oxford, 1993.

15. Organisation for Economic Co-operation and Development - OECD, *Education at a Glance 2012: OECD Indicators*, OECD Publishing, 2012.

16. Organisation for Economic Co-operation and Development - OECD, *World Energy Outlook 2011 Fact Sheet: global energy trends*, 2011.

17. Pereira, N. M., Fundos setoriais: avaliação das estratégias de implementação e gestão, *White paper n. 1.136*, Brasília, 2005.

18. Pereira, N. M., Sector Funds: evaluation of implementation and management strategies (in Portuguese), White paper n. 1.136, Brasília, 2005. Retrieved from: http://www.ipea.gov.br/pub/ td/2005/td_1136.pdf, [Accessed: 15-Feb-2007]

19. Pereira, N. M., and Simone P. F. Experiences to support sector technological innovation (in Portuguese), Journal of Technology Management & Innovation, Vol. 1, No. 3, pp 74-80, 2006.

8

Renewable Energy Investment in Emerging Markets: Evaluating Improvements to the Clean Development Mechanism

*Amy Tang[1], John E. Taylor[*2]*
[1]Civil Engineering and Engineering Mechanics Department, Columbia University, New York, USA

[2]Charles E. Via, Jr. Department of Civil and Environmental Engineering, Virginia Tech, Blacksburg, USA
e-mail: jet@vt.edu

ABSTRACT

In the past, industrialized countries have invested in or financed numerous renewable energy projects in developing countries, primarily through the Clean Development Mechanism (CDM) of the Kyoto Protocol. However, critics have pointed to its bureaucratic structure, problems with additionality and distorted credit prices as ill-equipped to streamline renewable energy investment. In this paper, we simulate the impact of policy on investment decisions on whether or not to invest in wind energy infrastructure in India, Brazil and China. Data from 2,578 past projects as well as literature on investor behaviour is used to inform the model structure and parameters. Our results show that the CDM acts differently in each country and reveal that while streamlining the approval process and reconsidering additionality can lead to non-trivial increase in total investment, stabilizing policy and decreasing investment risk will do the most to spur investment.

KEYWORDS

Agent-based modelling, Clean Development Mechanism, Energy finance, Renewable energy, Simulation, Wind power

INTRODUCTION

Over the next two decades, developing countries will account for 80% of increased electricity demand worldwide. Renewable energy is expected to play a large role in meeting this demand. However, development of renewable energy infrastructure has been hindered by lack of financing and difficulty attracting capital [1, 2]. The high risk perception leads to a higher cost of capital [3], and in developing countries, other social priorities act as competition for scarce funds [1]. Many countries also experience a "carbon lock-in" of incumbent energy sources, with existing infrastructure unable to adapt beyond fossil-fuel based resources [4].

In the past, industrialized countries have provided much of the investment and financing to develop renewable energy in emerging markets. This was due in part to the Clean Development Mechanism (CDM), a flexibility mechanism defined in the Kyoto Protocol aimed to reduce overall global greenhouse gas emissions by providing financial incentives for using zero or low-emitting technologies. Although many projects have been realized through the CDM, academics and industry professionals alike have deemed it inefficient and bureaucratic. Since the renewable energy industry is highly influenced

---

[*] Corresponding author

by regulatory drivers as compared to other industries [1, 4], it is especially important that policies encouraging renewable energy development are well-designed and effective. Given that future financing for renewable energy in developing countries will likely come from developed countries [3], the research community needs to evaluate the CDM as a tool to encourage such investment.

In this paper, we evaluate the impact of potential improvements to the CDM on investment using agent-based simulation techniques. We focus our analysis on the level of wind energy investment in India, Brazil and China under three different policy improvements. The next section provides details of the CDM including its most common criticisms and suggested improvements, as well as policy backgrounds for the three countries studied. This is followed by an introduction to agent-based simulation and details of our model. We then share our results and analysis before concluding with policy recommendations and a discussion of opportunities for future research.

KYOTO PROTOCOL AND THE CLEAN DEVELOPMENT MECHANISM

Program Description

The Kyoto Protocol is a binding pledge adopted by several industrialized ("Annex I") countries to reduce greenhouse gas emissions, in part by using zero- or low-emitting technologies for electricity production. Emission reduction is awarded with certificates that can be traded between countries, a system intended to reduce emissions in the most cost effective manner. The CDM is a flexibility mechanism defined in the Kyoto Protocol that allows projects in developing countries with no binding commitments to qualify for emission avoidance certificates, also known as Certified Emission Reduction (CER) certificates. The certificates are issued based on a baseline emission scenario and the avoided emission attributed to development of renewable energy projects. Projects must meet the additionality criteria, which requires evidence that the project would otherwise not be built without the added benefit of receiving credits. CERs can be used by Annex I countries to meet part of their binding commitments.

In theory, the CDM should encourage bi-lateral development of renewable energy projects in developing countries, in which an industrialized country's investment in a project is repaid in part by CERs. Recently, unilateral projects developed solely by the non-Annex I host country have also emerged, with the CERs sold via various exchanges to a country looking to meet its own binding target. As of 1 April 2013, 6,660 projects have been registered worldwide with 2,337 projects under review. The CDM is one of the primary ways in which developed countries subsidize renewable energy infrastructure in developing countries [3, 5, 6].

Key criticisms of the Clean Development Mechanism

Despite its apparent success and the potential impact on air pollutants [7], the CDM has been widely criticized, leading some researchers to suggest that it be abandoned in favour of fiscal regulation specific to individual countries alongside binding emission reduction goals [6, 8]. Other academics believe correcting some key flaws will greatly enhance the CDM's effectiveness. Three main issues examined here are its bureaucratic process, the effectiveness of "additionality," and distorted credit prices due to uncertainty around credit supply.

Bureaucratic process. Each renewable energy project must apply for CDM status in order to receive CER credits through a lengthy, "unwieldy and opaque bureaucratic structure" [5 pg. 91]. The Samana wind farm in Gujarat, India, for example, was commissioned in 2009 and did not receive CDM status until February 2013. Project developers rank the risk of non-approval as a primary concern when developing renewable energy in emerging markets [5]. The high transaction costs of participating in the program, therefore, reduce its benefits. Lewis [6] and Dechezlepretre et al. [9] suggest that streamlining the approval process as well as approving a program with multiple projects would greatly strengthen the design of the CDM.

Additionality. For a project to receive CDM approval, it must meet the additionality criteria, which requires developers to show that the project could not have been built without the additional revenue stream from sale of CER credits. However, these calculations of financial feasibility are based on self-reported rates of return of developers [3, 4] and there is no standard or uniform way through which this is done. Evidence from China [6], Brazil and India [5] imply some projects receiving CDM approval were not "additional."

Distorted credit prices. Lastly, critics have cited distorted CER credit prices as a major drawback of the CDM. Aside from the market reactions to the recent global recession and European debt crisis, the particularities of the CDM have led to irrational market fluctuation of credit prices. Because of the lengthy and oftentimes opaque approval process, there is uncertainty around the number of credits that will be available on the market in the future, clouding price-change signals [8]. Additionally, because the number of credits issued for a project is relative to a theoretical business-as-usual baseline, credits earned for identical projects in different locations will vary [5], adding to the uncertainty around credit supply. The future of the CDM is also undetermined, since countries have yet to renew their reduction commitments. This increases risks for developers, raising the required rate of return to participate in renewable energy projects for some investors and possibly deterring others. Stabilization of policy with greater certainty in future credit supply could increase investor confidence and lead to additional investments.

Need for quantitative analyses of Clean Development Mechanism improvements. In short, the CDM in its current state is ill-equipped to streamline renewable energy investment in emerging markets [8]. Although suggestions have been made to improve the CDM, no research has quantitatively evaluated the effect of these changes on total investment in renewable energy infrastructure. Informing policy makers of which changes can bring about the greatest increase in investment is valuable in designing future policy and continuing to encourage renewable energy investment in emerging markets. As a step in this direction, we use agent-based simulation to measure the total investment in wind energy infrastructure in India, Brazil and China under improvements to the CDM.

Key participants of the Clean Development Mechanism

India, Brazil and China are among the top countries in terms of CDM participation. According to the UNEP Risø Centre, China and India are the most active countries in Asia making up 55.0% and 29.7%, respectively, of all Asian CDM projects. Brazil leads with 35% of all Latin American CDM projects. Specifically for wind power, China, India and Brazil have the most installed capacity of wind energy infrastructure out of all countries eligible to participate in the CDM as of 2012 [10]. Additionally, all three countries are in the top ten countries with the most installed capacity during 2012 [10]. Developers in India were early to take advantage of the CDM program for wind energy, with almost 40 projects submitting for approval in Q4 of 2005. Between Q2 2007 and Q2 2010, approximately 30 projects per quarter applied for CDM status. Starting at the end of 2010, between 40 and 60 projects were proposed until the end of 2012. Chinese developers started regularly submitting wind power projects in Q1 of 2008 at a steady pace of approximately 40 per quarter. Since then, there has been a significant increase per quarter leading to a peak of 185 projects being considered in Q2 of 2012. Brazil dabbled with participation in the CDM for their wind projects starting in early 2006, although they had previously developed biomass/biogas projects under the program. Wind power developers did not start consistently applying for CDM status until the end of 2010, reaching a peak of 23 projects in Q4 of 2011. Although progress has been made, each country's potential wind resource is still much greater than the existing infrastructure, as shown in Table 1, and will continue to increase as turbine technology advances.

Data on wind power potential and installed capacity in Table 1 show that India, Brazil and China are important players in the CDM program with a rich dataset of projects to study, and continue to be leaders in the growing global wind industry. Below are brief descriptions of the wind development and regulatory landscape in each country.

Table 1. Wind power potential and installed capacity in India, Brazil and China

Country	Potential	Installed Capacity in 2012 [10] (% of potential)
India	102 GW [10]	18.4 GW (18%)
Brazil	350 GW [10]	2.5 GW (0.71%)
China	2380 GW [11, 12]	75.3 GW (3.16%)

India.   In 1983, the Wind Energy Program was started by the Ministry of Non-conventional Energy Sources (MNES) (currently the Ministry of New and Renewable Energy (MNRE)) and had broad goals of assessing wind resources, building demonstration projects and creating industry-utility partnerships [13, 14]. Although an initial leader in the wind energy industry compared to other developing countries, India's lack of an integrated energy framework and a national mission dedicated to wind has stunted its growth and allowed countries such as China to soar past its progress [10]. Despite this, several incentives have helped wind energy grow considerably over the past two decades, and India currently ranks fifth in installed capacity worldwide. At the national level, a bundle of tax incentives, including accelerated depreciation, low tariffs on imported wind energy technology and reduced or exempt tax for income from power sales helped developers largely using balance sheet financing (as opposed to project financing). Additionally, the Indian Renewable Energy Development Agency (IREDA) was established in 1987 to provide financing to developers. It instilled confidence in the economic viability of wind power and paved the way for private banks to lend to developers. Several state governments have enacted preferential feed-in tariffs, which encourage wind power development by providing a higher rate to electricity produced from wind. Recently, some states have also implemented Renewable Purchase Obligations (RPO) which require a certain amount of power produced to come from renewable sources, although lack of enforcement have negated much of its intended effect. Looking forward, some challenges facing the wind industry in India include implementing an integrated framework and a national feed-in tariff program, as well as continued development of transmission and other support infrastructure.

Brazil.   Brazil has long been a leader in renewable energy due to its numerous hydroelectric power plants. In an effort to diversify its energy portfolio and to hedge against low power production during the dry season and droughts, Brazil has aggressively promoted wind as part of its primary energy mix. Although a recent entrant into the wind power industry, the country is characterized by strong winds that allow turbines to operate for a longer period of time than in many other countries, giving the technology operational and financial advantages [15]. In 2002, the government created the Program for Incentive of Alternative Electric Energy Sources (Proinfa), which consolidated several previous actions promoting alternative energy development and provides subsidies and financial incentives funded through an electricity surcharge on power consumers [16]. The National Development Bank also provides financing for machinery and equipment through its subsidiary Finame. It has created a strong domestic industry and eleven international equipment manufacturers have opened production facilities in the country [10]. Starting in 2009, the government has held several wind-only energy auctions, essentially guaranteeing that wind energy operators received over half of the contracts to sell power in 2011 and 2012 [17]. However, the prices at which developers have agreed to sell electricity are extremely low and have caused concerns over the sustainability of future wind power growth [17]. In order to continue growth of its wind industry in the upcoming years, Brazil must ensure sufficient transmission infrastructure to keep up with the increase in electricity production and reduce financial risks through improved government regulations [10, 18].

China. Over the past decade, China has experienced rapid growth and currently has the highest installed capacity of wind energy infrastructure globally. In 2005, the government created the Renewable Energy Law of the People's Republic of China. After its adoption, a series of policies have been enacted to promote renewable energy development, including a dedicated renewable energy development fund supported by an electricity surcharge on consumers [12]. Additionally, feed-in tariffs have been used to give wind power a financial advantage. The manufacturing industry has also flourished due to policy incentives, creating a full supply chain with 88% of the domestic market occupied by Chinese manufacturers [12]. Since most developers are stated-owned enterprises supported by government-backed commercial banks eager to invest in infrastructure, the wind industry in China experiences lower financial risks and was largely unaffected by the 2009 crisis that upset the US and European industries [10]. Several obstacles still remain, however, most notably the inadequate transmission system and lack of regulations for wind power integration into the grid [10, 12].

Opportunities created by India, Brazil and China analyses. By analysing these three countries, not only can we gain insight into which improvements to the CDM are the best at encouraging development, but under what circumstances. By doing so quantitatively, we aim to measure the effect of improvements to the CDM policy will have on investment decisions. Our results will better inform policy makers as to where they should concentrate their efforts to create the greatest increase in total renewable energy investment in emerging markets.

AGENT-BASED SIMULATION

In order to determine the impact of changes to the CDM, we built an agent-based model to simulate the total investment in renewable energy given certain policy environments. Agent-based simulation was selected because of its extensive use in past studies to model and evaluate investment decisions under various policy scenarios. Mueller and de Haan [19] and Eppstein et al. [20] used agent-based simulation to determine how much incentives affect car purchase decisions and market penetration of plug-in hybrid electric vehicles. Jackson [21] analysed energy efficiency of a smart grid program and Veit et al. [22] determined the implications of transmission constraints on the German electricity market, both using agent based simulation.

Our agent-based model simulated investment decisions of an individual or firm seeking to develop a wind energy project. The decision is based on several factors that influence a project's potential revenue and profitability, including project properties, local conditions, and an investor's own characteristics. The model aggregated all investment decisions of investors over time to determine the total investment. Simulations were performed for investors in India, Brazil and China, allowing us to quantify the incremental investment on wind energy infrastructure under the presence of the CDM in each country. Sensitivity analysis was done to measure the impact of three improvements to the CDM program: streamlining the approval process, reconsidering the requirement of additionality, and reducing investor risk by stabilizing policy. The results allowed us to compare the effectiveness of the CDM and various program changes within and across countries.

Model structure

The agent-based model was designed and coded in AnyLogic, a powerful and robust simulation environment. The simulation environment is populated with agents representing individuals or firms, which we call "investors". An investor decides whether or not to build a wind energy project by evaluating its profitability. The decision process is represented in Figure 1.

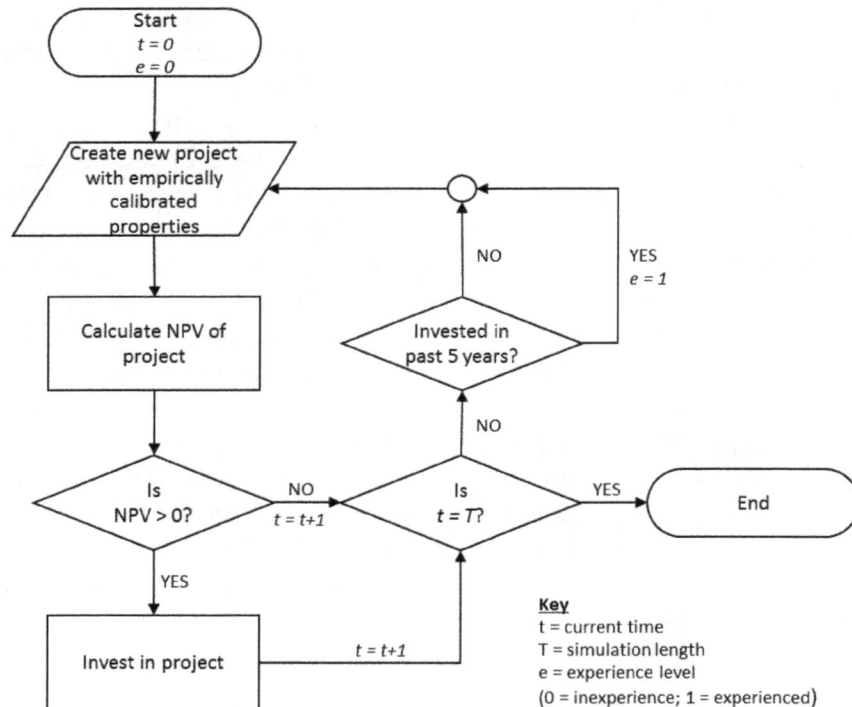

Figure 1. Decision process of an investor

By default, the simulation initially treats all investors as "inexperienced." Each year, an investor decides makes his or her decision by calculating the project's net present value (NPV) using its required rate of return. The cost and revenue streams for a project are based on four components: a) initial cost, b) yearly revenue from electricity sales, c) yearly operations and maintenance cost, and d) yearly revenue from Certified Emission Reduction (CER) credit sales. The NPV for each project is calculated as follows:

$$NPV = -\text{initial cost} + PV(\text{electricity revenue}) - PV(\text{O \& M}) + PV(\text{CER revenue})$$

where PV represents the present value of all future revenue or cost streams discounted to present day. The parameters used to calculate each of the four components and the discount rate used vary by country and are detailed in the next section. If the NPV is positive, then the project is built and the investor becomes "experienced". If an experienced investor does not build any projects for five years, then it reverts back to inexperienced. We ran the simulation for a period of T years and aggregated the costs of built projects to determine the total investment.

Baseline calibration

While the structure of the model remained the same for all three countries analysed, model parameters necessary to calculate costs and revenue were derived for each country using a database of all projects that have applied for CDM registration as of 1 April 2013, publicly available from the UNEP Risø Centre. After streamlining the data to focus on wind energy and removing duplicate projects that were listed more than once because of resubmissions, 83 projects were used to calibrate parameters for Brazil's model, 1544 for China's, and 931 for India's. For each country, a project's properties and situational characteristics are presented in Table 2. The rates of return used for experienced and inexperienced investors are based on a study done by Donovan and Nunez [3] on the cost of capital for renewable energy projects in emerging markets, and are also listed in Table 2.

Table 2. Parameters used in baseline agent-based simulation

	Parameter	India	Brazil	China
Environment	Number of investors	50	50	50
	Delay in receiving credits*	3 years	2 years	2 years
	Probability of getting CDM approval*	80.57%	77.5%	97.91%
Investor	Experienced investor discount rate**	19.06%	13.68%	11.08%
	Inexperienced investor discount rate**	14.80%	12.09%	8.32%
	Time of inactivity to revert back to inexperienced	5 years	5 years	5 years
Project	Years credits are received*	10 years	7 years	7 years
	Years electricity is produced	20 years	20 years	20 years
	Project size*	Exponential distribution with mean = 17.02 MW	Exponential distribution with mean = 75.13 MW	Either 50 MW or 200 MW
	Capacity factor*	23.3%	42.7%	24.5%
	Average CER credit received per MW capacity*	N (1891, 298)	N (1259, 409)	N (2032, 304)
	Average project cost per MW capacity*	N (122901, 237711) USD	N (1971742, 682573) USD	N (1289005, 160466) USD
	Yearly O&M cost per MWh electricity produced***	10 USD	10 USD	10 USD

* Empirical values derived from CDM project database
** Empirical values from [3]
*** Empirical values from International Renewable Energy Agency [23]

The initial cost of a project was calculated by multiplying the project size in megawatts [MW] of installed capacity by the average cost per MW. For each country, project size was randomly simulated based on an empirically-determined distribution of past projects. In the models for both India and Brazil, an exponential distribution with positive skewness (long right tail) was observed for project size with a mean of 17.02 MW and 75.13 MW, respectively. In China, the majority of CDM wind energy projects were between 40 and 50 MW, with an astonishing 999 (64.7%) projects with the exact size of 49.5 MW. This is because projects larger than 50 MW require approval from the National Development and Reform Commission (NDRC) while smaller projects are approved by local provincial governments and recorded with central government authorities [12], suggesting that developers prefer working with local authorities. Therefore in China's model, a project's size was simulated as either 50 MW with a probability of 0.9 or 200 MW with probability 0.1 in order to include the larger scale projects. The average unit cost per MW was also determined from information in the database. Project cost in all three countries followed a normal distribution, with no significant correlation found between project size and unit cost.

Yearly revenue from electricity sales was forecasted by multiplying the yearly electricity production by the expected price of electricity. It was received for the lifecycle of the project, which is 20 years. Yearly electricity production was calculated using capacity factors derived from the CDM project database for each country. The expected price of electricity was determined using feed-in tariff data for wind power in India [24] and China [12]. To maintain flexibility in the model, the feed-in tariffs for each Indian state were averaged and applied uniformly to all projects in India. Similarly for China, feed-in tariffs for individual zones were also averaged. A breakdown of the feed-in tariffs for each state and the average tariff utilized in our model is provided in Table 3 and Table 4. Brazil does not currently utilize feed-in tariffs for wind energy. Instead, the average price of the 2011 wind power auction of 99.58 BRL per MWh as reported by Bloomberg News and Merco Press was used to inform the model.

Table 3. Feed-in tariffs for electricity produced from wind power in India

State	Feed-in tariff per kWh [INR]
Andhra Pradesh	4.70
Gujarat	4.23
Haryana	Wind Zone 1 - 6.14
	Wind Zone 2 - 4.91
	Wind Zone 3 - 4.09
	Wind Zone 4 - 3.84
Karnataka	3.70
Kerala	3.64
Madhya Pradesh	4.35
Maharashtra	Wind Zone 1 - 5.67
	Wind Zone 2 - 4.93
	Wind Zone 3 - 4.20
	Wind Zone 4 - 3.78
Orissa	5.31
Average	4.448*

Source: [24]

*Converted to USD using ave. monthly exchange rate for 2011 of 1 USD = 44.899 INR

Table 4. Feed-in tariffs for electricity produced from wind power in China

Resource Zone	Feed-in tariff per kWh [CNY]
Category 1	0.51
Category 2	0.54
Category 3	0.58
Category 4	0.61
Average	0.56*

Source: [12]

*Converted to USD using ave. monthly exchange rate for 2011 of 1 USD = 6.464 CNY

While data for yearly operations and maintenance costs is not widely available, a survey of over 60 projects built in the 2000s revealed an average operations and maintenance cost of 10 USD per MWh of electricity produced (International Renewable Energy Agency [23]), making the total annual operations and maintenance dependent on project size. The unit cost was used for the models of all three countries and incurred for the lifecycle of the project.

Similar to revenue received from electricity sales, the revenue received from sale of CER credits was forecasted by multiplying the total number of credits by its expected price each year for which credits are received. Since the number of credits is calculated based on comparison with a theoretical business-as-usual baseline and may differ for projects of the same size, it was randomly generated in our models based on an empirically-determined distribution of values from the CDM project database for each country. The expected CER credit prices were simulated in MATLAB using techniques from [25]. We assume that daily returns follow Geometric Brownian motion, with the parameters calibrated from historical price data from 1 September 2009 to 31 August 2010. The starting CER price used for this simulation was the price on 1 September 2010: 13.44 EUR or 10.60 USD using the exchange rate of 1 EUR = 1.27 USD at that time. Expected future prices can move in either direction depending on a variety of factors. For example, using the techniques from [25] and running 10,000 simulations, the range of prices after 1 year (approximately 250 trading days) was between 3.93 EUR (31.58 USD) to 40.23 EUR (3.10 USD). The CDM allows a single 10-year crediting period or 7-year crediting period which can be renewed. For model simplicity and due to uncertainty surrounding credit renewal, a single value is used for each country's model. In India, the large majority of past projects applied for a 10-year crediting period while projects in both Brazil and China opted for a 7-year crediting period.

Two additional factors were taken into account: 1) probability of project acceptance and 2) delay in receiving credits. The database indicated that, on average, 80.47%, 77.5% and 97.91% of wind energy projects in India, Brazil and China, respectively, were accepted while the others did not receive CDM status. In our simulation, only projects that are approved were built. The database also indicated that there was an average delay of three years until credits are received in India and an average delay of two years in Brazil and China. In the models, this translated into a delay in receiving credit sale revenue which decreases the present value of credit sale revenue. Each investor evaluated each project independently. The initial costs of all built projects were aggregated to produce the total investment in wind energy infrastructure over the simulation time period.

Policy improvements and sensitivity analysis

The first improvement tested is streamlining the approval process for CDM registration and subsequent distribution of CER credits. To capture its effects, the value for delay in receiving CER credits was manipulated. A more efficient process equates to less waiting time to receive credits. Values of zero to five years were used to simulate total investment, holding all other parameters constant.

The second improvement tested is relaxing or eliminating the condition of additionality. In the models, this was equivalent to changing the probability that a project is accepted, with a probability of 1 meaning that "additionality" is excluded from the CDM. Values ranging from 0.6 to 1 in increments of 0.05 were tested holding all other parameters constant. Analysis was not performed for China since 97.91% of projects were already approved.

The last improvement tested is to reduce the magnitude of credit price distortion by stabilizing policy and providing greater clarity on future credit supply. Policy risk is captured in the discount rate of the investors; the more stable the policy, the lower the discount rate. Since there are different discount rates for experienced and inexperienced investors, this is done in terms of change to the discount rate. For example, "+2.5%" indicates an increase of 2.5% to both discount rates. Values of -5.0% to +5.0% in increments of 0.5% were tested holding all other parameters constant.

RESULTS

Baseline simulations were performed for T equal to 10 years, 15 years, and 20 years for each country. Thirty simulations were executed for each time period and the results for each simulation were averaged to obtain an average total investment for each value of T. Since the three countries differ greatly in population and electricity demand, the raw values of total investment are not appropriate for comparisons across countries. Instead, we simulated one more scenario in which we measure total investment in the absence of the CDM by removing the revenue stream from CER credit sales when calculating a project's NPV. Table 5 shows the results for India, Brazil and China and allows us to compare the impact of the CDM in each country. Sensitivity analyses were performed with T equal to 10 years. The results are detailed for each country in the following subsections. Again, 30 simulations were executed for each change of the appropriate parameter and then averaged.

Table 5. Total investment (average of 30 simulations) of the "No CDM" and "CDM baseline" scenarios for each country and each value of T

	Total investment with no CDM [billions USD]			Total investment with baseline CDM (as % of no CDM)		
T	10	15	20	10	15	20
India	1.78	2.89	4.18	2.38 (133%)	3.98 (138%)	5.69 (136%)
Brazil	6.63	9.80	12.8	6.01 (91%)	9.25 (94%)	12.4 (90%)
China	40.0	60.1	79.8	43.3 (108%)	63.8 (105%)	84.0 (105%)

India

The results from Table 5 show total investment of wind energy projects in India increased under the presence of the CDM, indicating that previously unprofitable projects became profitable with additional revenue from CER credit sales and were developed. Figure 2 further demonstrates this point by providing histograms of the NPVs of all wind energy projects considered over 10 years in two sample simulations, one including the CDM and one excluding it. More projects had positive NPVs during the simulation with CDM. However, the majority of projects still had NPVs of less than zero, which indicates that they were rejected by the investor and not built, even with the extra revenue stream.

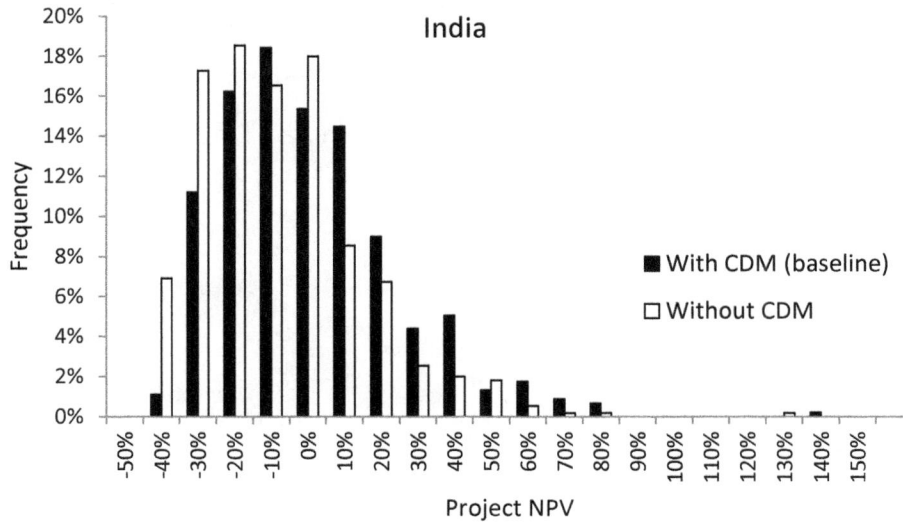

Figure 2. Histogram of *NPV* of all Indian projects considered in a sample simulations of *T* = 10 years

Figure 3 shows the impact on total investment when the delay in receiving CER credits was changed as compared to the baseline CDM scenario. The solid black square indicates total investment under the baseline scenario with a wait time of three years. As expected, total investment increased when the delay is shortened, with a linear trend. Under the best case scenario with no delay, total investment grew by over 30%.

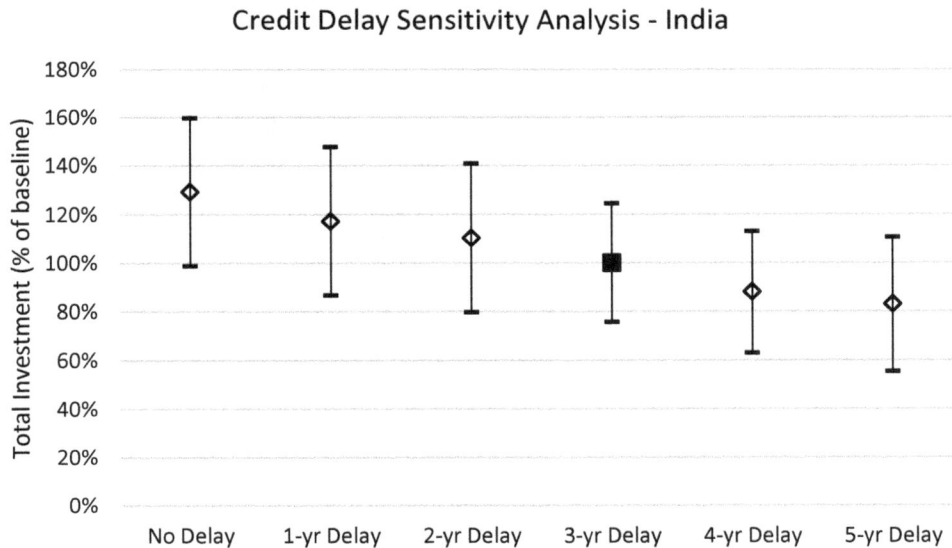

Figure 3. Sensitivity analysis for the delay of issuance of CER credits in India

To test the effect of relaxing or removing the requirement of additionality, we changed the probability of projects being approved and receiving CER credits. If the condition of additionality is completely removed, this is equivalent to the probability equalling 1. Figure 4 shows total investment compared to the baseline scenario. The solid black square represents total investment under the baseline simulation with a probability of 80.57%. As expected, total investment increased when the chances of receiving CER credits were increased. Under the best case scenario with guaranteed credits, total investment grew by approximately 35%.

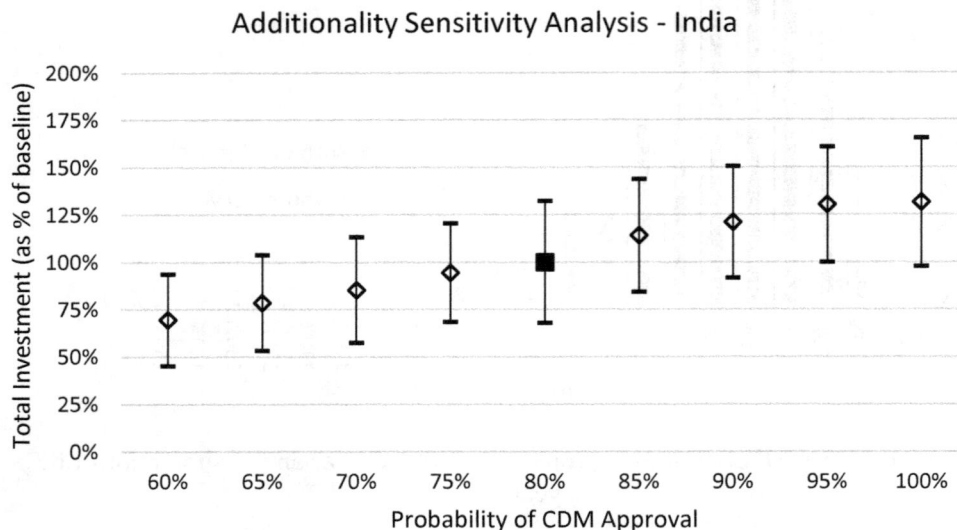

Figure 4. Sensitivity analysis for the probability of CDM program approval in India

Lastly, we used investor discount rate as a proxy for the policy stability, with a decrease in the discount rate representing an increase in stability. Figure 5 shows the total investment compared to the baseline when changing discount rate for both experienced and inexperienced investors. The solid black square represents total investment under the baseline scenario with no changes to either discount rate. Unlike the previous two policy improvements, there appears to be a nearly quadratic relationship between total investment and changes to the discount rate. Additionally, the impact on total investment was greater, reaching 343% of baseline total investment if the discount rate decreased by 5%.

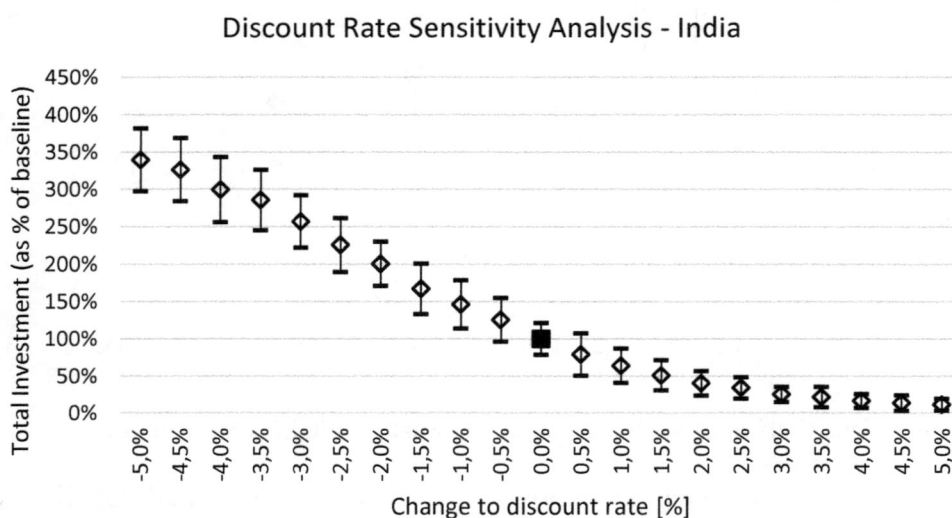

Figure 5. Sensitivity analysis of policy stability (using discount rate as a proxy) in India

Brazil

For wind energy projects in Brazil, Table 5 shows less investment when the CDM existed as compared to investment without the program. Although counterintuitive, this can happen when profitable projects that are built under the no CDM scenario are not built under the CDM scenario because they did not receive CDM approval. Discussion of this phenomenon can be found in the following section. Figure 6 displays histograms of the *NPV*s of all wind energy projects considered over 10 years in two sample simulations, one which included the CDM and one which excluded it. Unlike the histogram of project *NPV*s in India, the aggregate of projects with negative *NPV*s was approximately the same under both scenarios. In both cases, the majority of projects still had *NPV*s of less than zero and were not built.

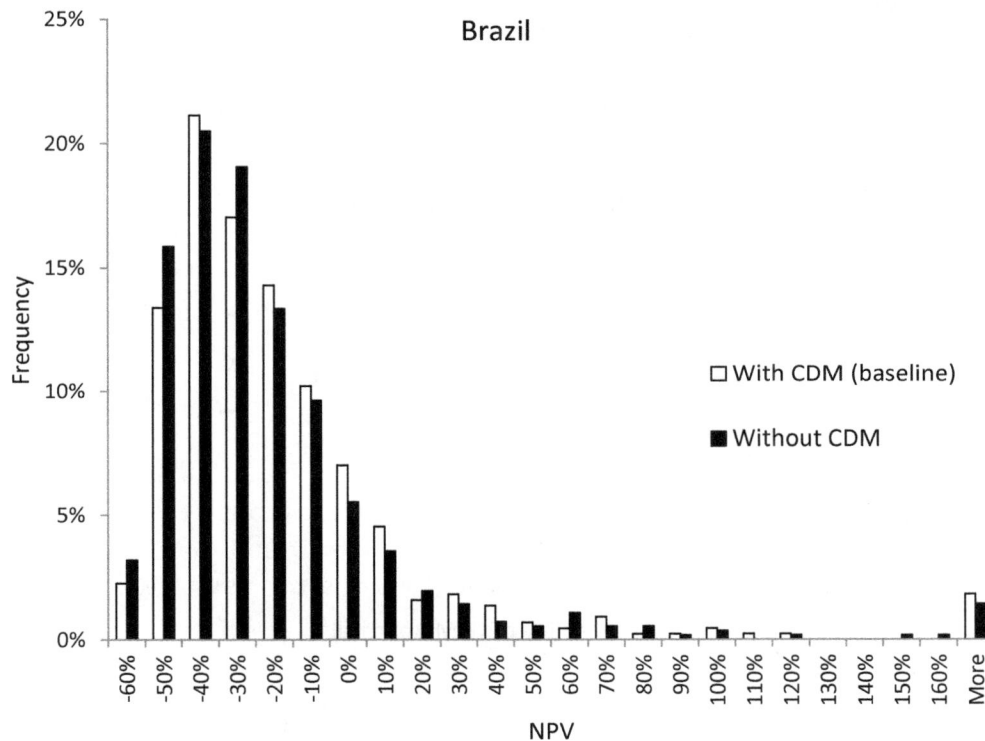

Figure 6. Histogram of *NPV* of all Brazilian projects considered in a sample simulations of $T = 10$ years

Figures 7, 8 and 9 show the results of scenario analysis for Brazil of streamlining the approval process, reconsidering additionality, and changes to the discount rate as a result of policy stability, respectively. The solid black squares in each figure represent total investment under the baseline CDM scenario and the top and bottom bars indicate two standard deviations of investment over 30 simulations. Streamlining the approval process and allowing investors to receive CER credits earlier had an almost negligible effect on total investment. Increasing the number of projects receiving CDM status does lead to additional investment, with an increase of 35% of all projects are approved. Similar to the results of the India analysis, stabilizing policy and decreasing investor discount rates drastically increased total investment.

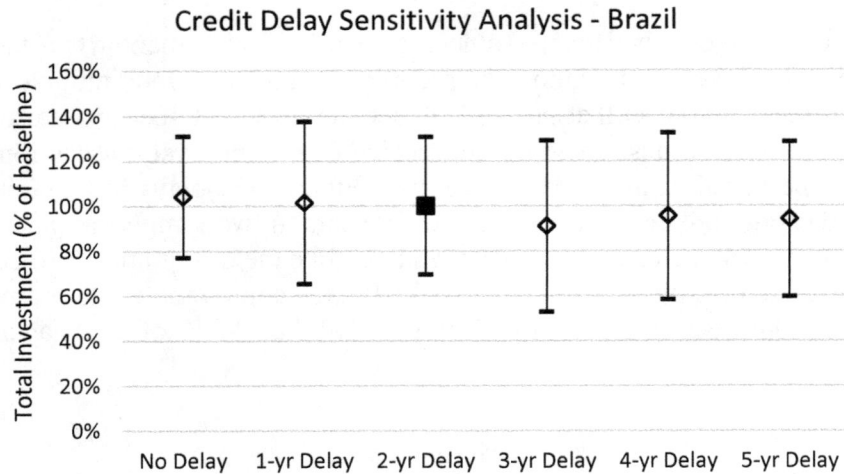

Figure 7. Sensitivity analysis for the delay of issuance of CER credits in Brazil

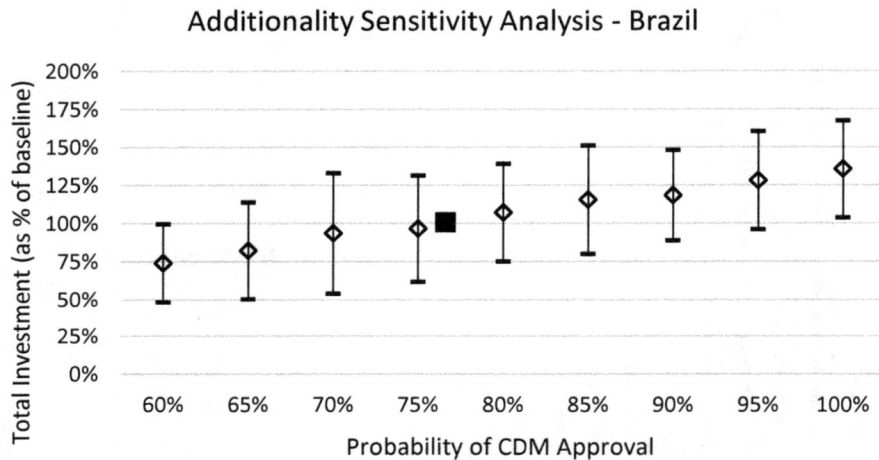

Figure 8. Sensitivity analysis for the probability of CDM program approval in Brazil

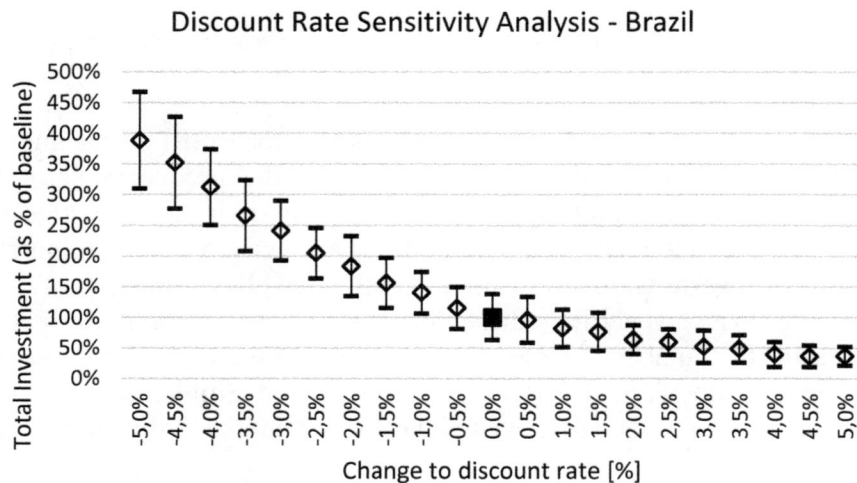

Figure 9. Sensitivity analysis of policy stability (using discount rate as a proxy) in Brazil

China

The presence of the CDM in China increased total investment in wind energy infrastructure according to the results of Table 5. However, most projects are profitable even without the extra revenue of CER credit sales, as shown in Figure 10. Therefore we can expect that any improvements to the CDM will not drastically increase investment, reflected in Figures 11 and 12. Because over 97% of projects are already receiving CDM project approval, sensitivity analysis was not performed on this parameter. Same as with the results of the previous two countries, the solid black squares represent baseline CDM investment with the top and bottom bars indicating two standard deviations of investment over 30 simulations.

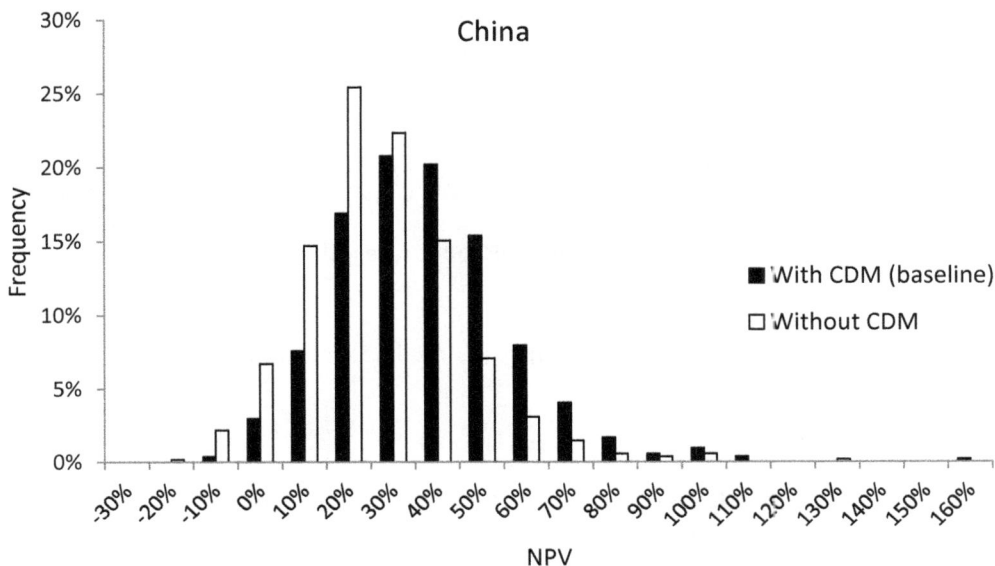

Figure 10. Histogram of *NPV* of all Chinese projects considered in a sample simulations of $T = 10$ years

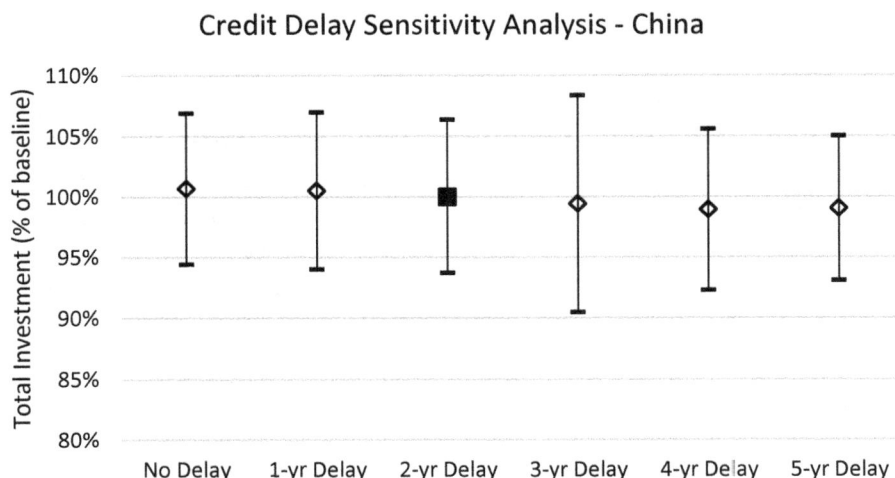

Figure 11. Sensitivity analysis for the delay of issuance of CER credits in China

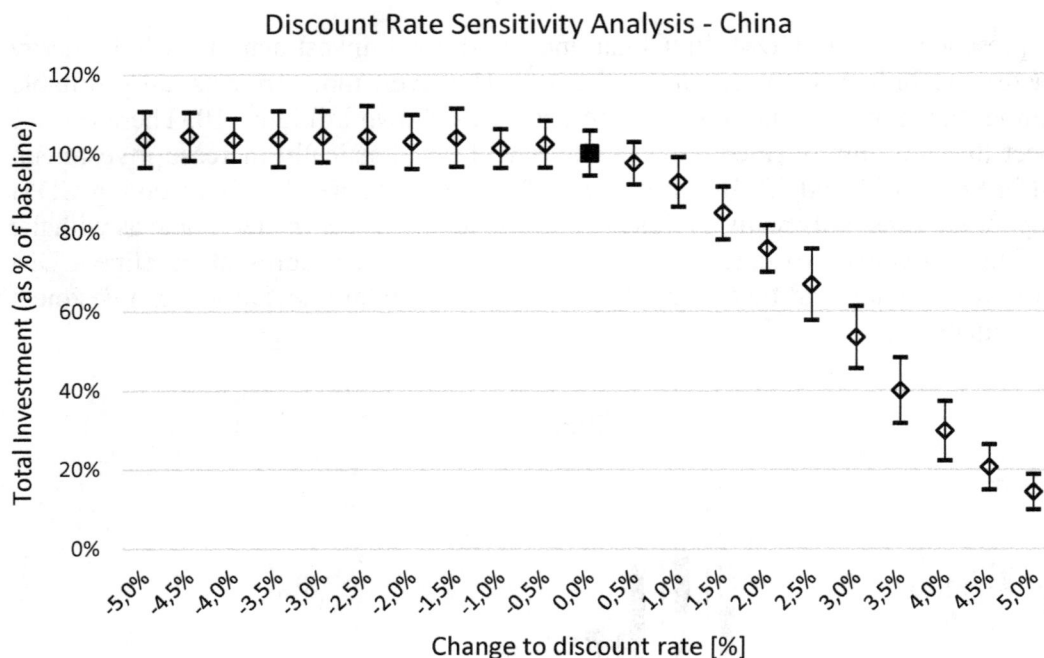

Figure 12. Sensitivity analysis of policy stability (using discount rate as a proxy) in China

DISCUSSION

Although originally intended as a reference point for scenario analysis of the three policy improvements, the CDM baseline results provide some interesting insights for individual countries and comparison across them. We discuss below two key findings: 1) why most projects considered in China have positive *NPV*s while the reverse is true for India and Brazil; and 2) why Brazil's level of investment actually decreases under the presence of the CDM. We then discuss the results of the policy improvements to the CDM and its implications for the future of the program as well as other energy policy measures.

Baseline comparisons between countries

In the sample simulations of all wind energy projects considered, Figures 2, 6 and 10 show that the *NPV*s of most projects in India and Brazil are negative and rejected, while most projects in China are profitable and built. Since this is the case for the scenario when the CDM does not exist, we can attribute China's superior performance to the landscape in which projects are developed. A study done by [26] showed the importance of a comprehensive strategic approach to initiatives used to promote diffusion of renewable energy technologies. We use this strategic structure matrix as a framework to analyse each country's existing policies.

The matrix breaks down each policy into five different categories in terms of which area it targets. Table 6 summarizes the five categories, and details on development and usage of the matrix can be found in [26]. Table 7 classifies the policies and incentives of each country as discussed in the "Key participants of the Clean Development Mechanism" section of this paper.

Table 6. Strategic Structure Matrix [26]

Lowering a Diffusion Barrier	Empowering Actors	Creating an Enabling Environment	Direct or Indirect Influence	Supply or Demand
Technical: decreases uncertainty with technical capabilities - or - Financial: decreases cost or increases future value - or - Regulatory: decreases regulatory hurdles to diffusion	R&D: increases support of research laboratories and organizations developing new technologies - or - Systems and Infrastructure: provides support for surrounding infrastructure (physical, market, etc.) - or - Knowledge and learning: strengthens ties between facilitating organizations	Institutional Change: improves institutional landscape for easier diffusion of the technology - or - New/Expanded Market: creates a new market or expands existing market - or - Advocacy coalition: creates or promotes pro-diffusion organizations	Direct: targeted primary at technology being diffused - or - Indirect: targeted primary at infrastructure surrounding the technology	Supply: ultimately decreases costs or lowers barriers associated with adoption of the technology - or - Demand: ultimately raises the payoff after adoption of the technology

Table 7. Classification of country specific policies and incentives

	Initiative	Barrier	Actor	Environment	Direct/ Indirect	Supply/ Demand
India	Bundle of Tax incentives	Financial	Systems and infrastructure	Market	Direct	Demand
	Loans from IREDA	Financial	Systems and infrastructure	Market	Indirect	Supply
	Feed-in tariffs	Financial	Systems and infrastructure	Market	Direct	Demand
China	Renewable Energy Law	Regulatory	Knowledge and learning	Institutional change	Direct	Supply
	Government-backed banks	Regulatory	Knowledge and learning	Advocacy coalition	Indirect	Supply
	Renewable Energy Development Fund	Financial	Systems and infrastructure	Market	Direct	Supply
	Feed-in tariffs	Financial	Systems and infrastructure	Market	Direct	Demand
Brazil	Proinfa	Financial	Systems and infrastructure	Market	Direct	Supply
	Financing from Finame	Financial	Systems and infrastructure	Market	Indirect	Supply
	Wind Auctions	Financial	Systems and infrastructure	Market	Direct	Demand

China's policies toward wind power development are more comprehensive in terms of the number of strategies covered. Most notably, its renewable energy law and government involvement in the commercial banking sector is able to overcome regulatory hurdles to diffusion, strengthen ties between facilitating organizations, improve the institutional landscape for wind power deployment, and advance organizations interested in developing more renewable energy infrastructure. These strategic areas are missing from the bundle of policies and incentives offered by India and Brazil, and support previous findings that a comprehensive approach to increasing diffusion is most effective [26]. By doing so, China offered investors stability and reduced risk. It was thought of as a "safe haven" for wind power investment when the effects of the global financial crisis reached the wind industry in other parts of the world [10], explaining the low required rates of return found by Donovan and Nunez [3] and used as the discount rates in our simulation. The importance of the discount rate to total investment is clear when looking at Figures 5, 9 and 12, which will be discussed in more detail in the following subsection.

Brazil and India's focus on market-forming financial incentives are usually found in the early phases of technology diffusion. While this is understandable for Brazil since it recently entered the market, India has been a part of the wind power industry for over 20 years. India's lack of a comprehensive national policy for wind power combined with the segmented nature of the energy sector emphasizing states is hurting its growth. Researchers agree that a unified vision with national goals is essential for India's continued growth [10]. As part of a study done by Martins and Pereira [18], questionnaires were sent to companies, academic and research institutions, and national organizations and associations asking them to rank initiatives to expand renewable energy deployment in Brazil. Two of the highest ranked issues were a) improve government regulations and b) reduce financial risks. For both countries, the sustained regulatory uncertainty means investors will continue to require higher rates of return.

Table 5 shows that total investment in wind energy infrastructure in Brazil actually decreased under the presence of the CDM. This seemingly counterintuitive finding is due to the country's ratio of revenue from electricity sales to CER credit sales. Brazil's wind resources allow increased power production, leading to an average capacity factor of 42.7% compared to 23.3% for India and 24.5% for China (see Table 2). This may be a contributing factor to the low bid prices from investors in recent auctions. However, average capital cost is much higher, leading to the highest frequency of non-profitable projects in our simulation. Unfortunately for Brazil, since CER credits issued through the CDM are based on comparison against a baseline scenario, Brazil's success with hydropower actually hurts its wind industry with respect to participation in the program. The average number of credits it receives for the same size of a wind farm is much less than that received by India or China since Brazil's energy mix is already heavily reliant on emission-free sources. Combined with the higher volume of electricity produced, it means that revenue received from electricity sales matters more than revenue from CER credit sales when calculating NPV. This explains the decrease in investment under the presence of the CDM. The extra credit revenue does little to push unprofitable projects into profitability, while rejection of approximately 33% of projects from the CDM means that they will not be built, even if they are profitable.

China's tight regulatory control over the wind power industry is rare and should not be considered normal for renewable energy development in emerging markets. While the feed-in tariff and CER credit sales generate extra revenue for the investors, some projects are still unprofitable, as presented in our simulation results from India and Brazil. This highlights the need for continued financial support of wind energy infrastructure in emerging economies. Although some projects may be financially feasible without additional government support, to achieve the level of investment necessary to meet future demand, incentives must continue to exist. The different reactions to the CDM from the three countries in our analysis, however, support the view that CDM should play a secondary role and developing countries should adopt their own commitments to reduce emissions [6, 8].

Policy Improvements

Several improvements to the CDM may encourage greater investment in renewable energy infrastructure in emerging markets. Since most of China's wind power projects were already being developed under the CDM baseline simulation, improvements to the program cannot meaningfully increase investment. It should be noted that these improvements may increase the profit margin for individual projects and investors in China. However, the only way to increase total investment is to increase revenue for a proposed project with negative *NPV* so that it is profitable and subsequently built. Therefore we focus our analysis here on Brazil and India.

Our simulation results shows that streamlining the approval process and shortening the length of time it takes to receive CER credits raises the total level of investment in wind power in India. As a regulatory improvement that strengthens ties between organizations promoting diffusion of wind power, it is unsurprising for India to have such a positive reaction. However, it had a negligible effect on total investment for Brazilian projects. This is likely due to the fact that the revenue from credit sales is less of a factor in calculating *NPV* as discussed previously. Few unprofitable projects that are already receiving credits will become profitable when changing the present value of credit revenue. Allowing more projects to receive credits, however, does lead to higher investment.

Investments in both India and Brazil increased when the goals of additionality were reconsidered, with total investment rising by 31% and 35%, respectively. By allowing all renewable energy projects to receive credits regardless of their financial viability in its absence, previously unprofitable projects became profitable, were developed, and increased total investment. This is encouraging to researchers who believe the CDM would be greatly improved without the additionality criteria [3-5].

Most striking, however, is the effect of stabilizing policy and lowering investment risk for developers. A 1% decrease in discount rates leads to, on average, 50% more simulated projects becoming profitable in India and 56% more projects in Brazil, with a doubling of investment realized at a 2% decrease in India and a 2.5% decrease in Brazil. This has clear implications for policy makers. No matter how attractive they make the financial incentives, having them be stable (for example, longer periods until renewal) will do much more to increase investment. The baseline comparisons between the three countries also echo this fact, as the simulation results for China, the country with the lowest discount rates, show many more projects being realized and higher levels of investment. Additionally, the sensitivity analysis on the Chinese discount rates show investment cut in half when the discount rates increase by 3%, which is still lower than the discount rates used for the Brazil simulation. The importance of policy stability and lower risk has been observed historically as well as by many researchers. For example, the appearance and disappearance of tax credits in the US led to a bust-boom development cycle for wind power in the 1990s [27].

CONCLUSION

In an attempt to encourage renewable energy investment, the CDM has provided an incentive for parties from industrialized countries to develop renewable infrastructure in emerging markets. As the future of the CDM is currently being debated, there have been many suggestions to improve the program and address some of its shortcomings, including streamlining its bureaucratic process, relaxing or removing the requirement of additionality, and lowering investor risk by stabilizing policy. We utilized agent-based simulation to quantify total investment in three countries: India, Brazil and China, which are key participants in both the CDM and the global wind power industry. While we intended primarily to evaluate the effects of various regulatory improvements to the CDM, country comparisons of the baseline scenario provided valuable insights. The success of China's wind power development can be attributed in part to a comprehensive approach to renewable energy policy and initiatives, especially the high level of regulatory oversight. However, this should not be considered widespread and the results from India and Brazil reveal the need for continued financial incentives for emerging markets. Our results from the sensitivity analysis of policy improvements show that, compared with the baseline, streamlining the approval process and increasing the odds of project approval has the potential to add significant investment to the wind power sector. However, stabilizing policy is even more effective in increasing the level of investment. This provides key insights to policymakers when designing future policy to encourage renewable development in developing countries. Future research can expand this analysis to other countries that also participate in the CDM and incorporate simulation of investment decisions into larger models of policy evaluation.

ACKNOWLEDGMENTS

This material is based upon work supported by the National Science Foundation under Grant no. 1142379. Any opinions, findings, and conclusions or recommendations expressed in this material are those of the authors and do not necessarily reflect the views of the National Science Foundation.

REFERENCES

1. Gupta, S., Financing Renewable Energy, in Toth, F.L. (Ed.), Energy for Development: Resources, Technologies, Environment, *Environment and Policy*, Vol. 54, pp. 171-186, 2012.

2. Liming, H., Financing rural renewable energy: a comparison between China and India, *Renewable and Sustainable Reviews*, Vol. 13, pp. 1096-1103, 2009.

3. Donovan, C., Nunez, L., Figuring what's fair: the cost of equity capital for renewable energy in emerging markets, *Energy Policy*, Vol. 40, pp. 49-58, 2012.

4. Unruh, G. C., Carrillo-Hermosilla, J., Globalizing carbon lock-in, *Energy Policy*, Vol. 34, No.10, pp. 1185-1197, 2006.

5. Hultman, N. E., Pulver, S., Guimaraes, L., Deshmukh, Kane, J., Carbon market risks and rewards: firm perceptions of CDM investment decisions in Brazil and India, *Energy Policy*, Vol. 40, pp. 90-102, 2012.

6. Lewis, J. I., The evolving role of carbon finance in promoting renewable energy development in China, *Energy Policy*, Vol. 38, pp. 2875-2886, 2010.

7. Eto, R., Murata, A., Uchiyama, Y., Okajima, K., Assessment of co-benefits of Clean Development projects based on the project design documents of India's power's sector currently under registration and validation, *Journal of Sustainable Development of Energy, Water and Environment Systems*, Vol. 1(4), pp. 326-339, 2013.

8. Zavodov, K., Renewable energy investment and the clean development mechanism. *Energy Policy*, Vol. 40, pp. 81-89, 2012.

9. Dechezlepretre, A., Glachant, M., Meniere, Y., The Clean Development Mechanism and the international diffusion of technologies: an empirical study, *Energy Policy*, Vol. 36, pp. 1273-1283, 2008.

10. Global Wind Energy Council, *Global Wind Report: Annual Market Update 2012*, 2012a.

11. Phadke, A., Bharvirkar, R., Khangura, J., *Reassessing Wind Potential Estimates for India: Economic and Policy Implications*, Lawrence Berkley National Laboratory LBNL-5077E, 2011.

12. Jiang, L., Chi, Y., Qin, H., Pei, Z., Li, Q., Liu, M., Bai, J., Wang, W., Feng, S., Kong, W., Wang, Q., Wind Energy in China, *IEEE Power and Energy Magazine*, pp. 36-46

13. Bhattacharya, S. C., Jana, C., Renewable energy in India: historical developments and prospects, *Energy*, Vol. 34(8), pp. 981-991, 2009.

14. Pillai, I. R., Banerjee, R., Renewable energy in India: status and potential, *Energy* 34(8), pp. 970-980, 2009.

15. Simas, M., Pacca, S., Socio-economic benefits of wind power in Brazil, *Journal of Sustainable Development of Energy, Water and Environment Systems*, Vol. 1(1), pp. 27-40, 2013.

16. Dutra, R. M., Szklo, A. S., Incentive policies for promoting wind power production in Brazil: scenarios for the Alternative Energy Sources Incentive Program (PROINFA) under the new Brazilian electric power sector regulation, *Renewable Energy*, Vol. 33, pp. 65-76, 2008.

17. Leahy, J., *"Brazil: wind gathers force in mix of renewable sources"* Financial Times, 15 May 2013.

18. Martins, F. R., Pereira, E. B., Enhancing information for solar and wind energy technology deployment in Brazil, *Energy Policy*, Vol. 39, pp. 4378-4390, 2011.

19. Mueller, M. G., de Haan, P., How much do incentives affect car purchase? Agent-based microsimulation of consumer choice of new cars—Part I: Model structure, simulation of bounded rationality, and model validation, *Energy Policy*, Vol. 37, pp. 1072-1082, 2009.

20. Eppstein, M. J., Grover, D. K., Marshall, J. S., Rizzo, D. M., An agent-based model to study market penetration of plug-in hybrid electric vehicles, *Energy Policy*, Vol. 39, pp. 3789-3802, 2011.

21. Jackson, J., Improving energy efficiency and smart grid program analysis with agent-based end-use forecasting models, *Energy Policy*, Vol. 38, pp. 3771-3780, 2010.

22. Veit, D. J., Weidlich, A., Kraft, J. A., An agent-based analysis of the German electricity market with transmission capacity constraints, *Energy Policy*, Vol. 37, pp. 4132-4144, 2009.

23. International Renewable Energy Agency, *Renewable Energy Technologies: Cost Analysis Series*, IRENA Secretariat, Abu Dhabi, 2012.

24. Global Wind Energy Council, *India Wind Energy Outlook*, 2012b.

25. Tang, A., Chiara, N., Taylor, J. E., Financing renewable energy infrastructure: formation, pricing and impact of a carbon revenue bond, *Energy Policy*, Vol. 45, pp. 691-703, 2012.

26. Tang, A., Taylor, J.E., Mahalingam, A., Strategic Structure Matrix: A framework for explaining the impact of superstructure organizations on the diffusion of wind energy infrastructure, *Energy Policy*, Vol. 63, pp. 69-80, 2013.

27. Bird, L., Bolinger, M, Gagliano, T., Wiser, R, Brown, M., Parsons, B., Policies and market factors driving wind power development in the United States, *Energy Policy*, Vol. 33, No 11., pp. 1397-1407, 2005.

9

Energy Storage Needs in Interconnected Systems Using the Example of Germany and Austria

Thomas Weiss[1]*, Karl Zach[2], Detlef Schulz[1]
[1]Electrical Power Systems, Helmut Schmidt University, Hamburg, Germany
e-mail: thomas.weiss@hsu-hh.de
[2]Energy Economics Group, Technical University of Vienna, Austria

ABSTRACT

The share of renewable energies on the net electricity consumption is rising steadily. Especially intermittent, non-controllable sources like wind and sun are gaining importance. With an installed amount of non-controllable power that exceeds the yearly peak load, situations can occur with a surplus of energy in electricity supply systems. This surplus will rise strongly with the share of fluctuating renewable energies on the net electricity consumption. A lot of studies and experts come to the conclusion that energy storage will be needed to handle these surpluses. The questions that still have to be answered are when and how. Especially the German electricity system will have very high storage needs because of a very strong and fast development of wind and solar power. There are a lot of technologies and approaches to overcome this problem. However, Pumped Hydro Energy Storage (PHES) systems are up to date the most efficient and economic bulk energy storage technology. On the one side there are no high potentials in natural sites for the installation of PHES schemes in Germany. On the other side the Austrian PHES system has still a very high potential. Up so far, the prospects are used just to a small amount. Especially the seasonal Hydro Energy Storage (HES) still holds a very high potential. In this paper the combination of the Austrian and the German energy supply system will be investigated with respect to the development of renewable energies. The overall energy storage needs are evaluated for each country as well as in the combined system, taking also into account the development and the influence of the transmission system capacity between the two countries.

KEYWORDS

Energy storage, Energy storage needs, Renewable energies, Energy system modelling.

INTRODUCTION

When moving towards high shares of fluctuating, decentralized renewable energy units in an electricity system, there are some problems to overcome in the next decades. Two of the main challenges are the local and the seasonal compensation.

A major portion of electrical energy will not be generated in close distance to big urban load centres or suitable grid connection points anymore. Instead it will be located in regions where the best natural resources are, e.g. offshore wind farms, which can be far away from the big load consuming areas. For this problem of regional balancing of energy, grid extension as well as reinforcement will be essential. The other challenge will be that these units produce energy when there are the right weather conditions (e.g. wind or sun) and not exactly when the energy is consumed. Energy storage will sooner or later be needed to seasonally balance out the fluctuations of the energy output of renewable energies (RE) to

---

* Corresponding author

adapt the production to the demand. There are some interesting and promising solutions, like Demand Side Management (DSM) or Demand Side Response (DSR), to adapt the demand to a mainly volatile electricity production. These solutions can help reducing the energy storage needs but will not have the right capacity to avoid the need of additional energy storage. There is a broad agreement amongst experts that energy storage will be needed to a high extent, only the point in time and the amount sometimes still differs widely, e.g. [1-4].

When taking a look at energy storage systems (ESS) that are operating up to date, the only technology that is available concretely and in a relevant scale for bulk energy storage is the pumped hydro energy storage system (PHES). PHES normally requires adequate geological surrounding and resources. These natural resources do not appear in all countries. In Germany, e.g. adequate sites for PHES installations are rare and sometimes difficult to access due to environmental concerns and public opposition. There are some new and innovative approaches using artificial structures for the construction of PHES facilities. Two very interesting possibilities are the use of former open cast mining structures [5] or of the locks and canal lifts of the national waterways with the already existing pumps and reservoirs [6]. Nevertheless, even when taking the potentials of these approaches into account, there is still a very high need for energy storage until the year 2050. Austria on the other side still has a very high, unused potential for natural sites for PHES systems. Furthermore, the storage reservoirs of Seasonal Hydro Power Plants (SHPP) in Austria gather a very high potential when retrofitting it with pumps. Austrian PHES systems already participate on the German electricity market and, due to lower grid fees, are often economically more viable than the systems in Germany.

This paper carries out an approach of using the potentials in the Austrian Alps to overcome the energy storage needs not only of Austria but also to help overcome the ones of Germany.

METHODOLOGY

In this chapter the methodology of the computation algorithm will be explained and the scenarios under investigation are defined. The time resolution for this investigation is in hourly values and for the period of one year.

Modelling renewable energies and energy storage needs

For the modelling of the different renewable energy sources, weather data from national weather services as well as from the software Meteonorm® are used together with detailed technical models of photovoltaic (PV), concentrated solar power (CSP) and wind turbines. The models have been validated with real feed-in data of the particular renewable energy. With the generated feed-in curves from renewable sources, the residual load and the surpluses of energy can be calculated. The residual load is here defined as the load demand minus the non-controllable production of renewable energies. This includes wind (P_{Wind}) and solar power (P_{PV}) but also a part of hydropower (P_{HP}), mostly small non-controllable run-of-the-river plants. κ_{HP} represents the non-controllable factor for hydropower reaching from 0 (all hydropower plants are fully controllable) to 1 (none are controllable). Equation 1 is showing how the RL is calculated for each time step t.

$$RL_t = P_{d,t} - P_{Wind,t} - P_{PV,t} - \kappa_{HP} \times P_{HP,t} \tag{1}$$

After the calculation of the residual load the operational strategy of the ESS has to be defined. The operational strategy of an ESS can influence the outcomes of the final energy storage needs [7, 8]. In this paper the operational strategy is to integrate as much renewable energy as possible. To achieve this goal the ESS follows in general a simple peak-shaving valley-filling strategy, but with an important modification. If the residual load is expected to be negative, the ESS plans its operation before in a way, that enough or at least the maximum capacity is available to integrate the surplus of renewable energies. The operation planning is done with weather forecast models. To estimate the total energy storage needs, two ESS are integrated in the algorithm. ESS 1 contains the existing and planned facilities, in Germany it consists of 8 GW/60 GWh of PHES capacity and 321 MW/2.4 GWh Compressed Air Energy Storage (CAES) capacity. ESS 2 has an unlimited capacity and power and is not connected to a specific technology. Due to its unlimited nature and the optimization strategy of the algorithm, it is ensured that all energy from renewable sources can be integrated with minimum energy storage capacity. The finally used power and capacity of ESS 2 is then an indicator for the energy storage needs to fulfil the desired optimization target. In a next step, the total energy storage needs are separated into short and long term energy storage needs. This is done either with a discrete Fourier Transformation or with a floating average [9,10]. Short term means in this case full load hours of up to 24 hours and long term of multiple days to weeks. A detailed description of the modelling of the particular renewable energy sources and the energy storage needs, short as well as long term, can be found in previous work [9, 11].

Scenarios under investigation

Different development scenarios for renewable energies have been investigated for Austria as well as for Germany. The year of interest is 2050. The German government set the goal of renewable energy source electricity generation (RES-E) to 80% of the gross electricity consumption. To achieve this goal, the development of wind and solar power (mainly PV) will be a key element. For these two technologies 3 different cases representing 3 different development scenarios were defined, see Table 1. In scenario A, an equal development of both, wind and solar power, is assumed with the expected installed capacities of the particular National Renewable Energy Action Plan (NREAP) [12], scaled accordingly to reach 80%. Scenario B and C then represent a favoured development of wind and solar power respectively, see Table 1. The overall installed power of RES is highest in scenario C. This is due to the lower full load hours of PV compared to wind power. Thus a higher capacity is needed to produce the same amount of energy. For Austria only 2 scenarios are investigated for the year 2050, a Business As Usual (BAU) scenario and a GREEN scenario, named B and C respectively. The BAU scenario expects a low growth in RES-E but a strong increase in electricity consumption. The GREEN scenario expects just a low growth in electricity consumption and very strong development of PV. These two scenarios were derived with the GREEN-X model of the Energy Economics Group of the Technical University of Vienna [13].

For the combined system the GREEN scenario of Austria was combined with the 2 RE development scenario B and C of the German model. The installed capacities of the different RE-technologies in all scenarios are summarized in Table 1. In all scenarios the installed power of renewable energies exceeds the yearly peak load. Except of the Austrian scenario B, the installed amount of viable, non-controllable RE sources wind and photovoltaic (PV) even exceed the peak load more than twice. In the combined system of Germany and Austria (scenario CC), just the installed PV power is more than 30 GW higher than the highest load. The total installed power of RES in the combined system ranges from 182.5 GW to 220.6 GW.

Transmission capacities and technical constraints for energy storage systems

Nowadays (winter 2010/11) Germany and Austria have 2 grid connection points on high voltage level with a total transmission capacity of 2,200 MW. To be able to store the surplus of renewable electricity produced in Germany in the Austrian Alps, the transmission capacity would have to be increased. The influence of the transmission capacity can be seen in the next chapter.

Furthermore technical limitations and constrains will have to be set for the transmission capacities as well as the short and long term energy storage systems. For the energy storage systems, the power in charging as well as in discharging mode, the capacity factor and the cycle efficiencies can be adapted.

Table 1. Overview of the scenario assumptions for Germany in the years 2020 and 2050 [14, 4]

[MW]	DE - 2050 Scenarios			AT - 2050 Scenarios		Total AT-DE	
	A	B	C	B	C	BC	CC
Wind (onshore)	60,000	63,000	55,000	4,240	4,710	67,710	59,710
Wind (offshore)	26,000	30,000	21,000	0	0	30,000	21,000
PV	70,000	45,000	100,000	7,880	26,000	71,960	126,960
Hydropower							
Run-of-River		5,700		7,200		12,900	
PHES		8,000		3,600		3,600	
HES				9,200		17,200	
Small Hydropower				250		250	
Other RES-E	4,200	4,200	4,200	2,430	2,000	7,000	
Yearly Peak Load		79.1 GW		20.7 GW	13.6 GW	92.5 GW	92.5 GW
Energy Consumption [TWh]		~500		126	83	~583	~583
RES-E Generation [TWh]		~400		75	94	~494	~494
RES-E Share [†]	~80%	~80%	~80%	~60%	~110%		

RESULTS

The theoretical capacity of the Austrian reservoirs with connection to hydropower plants has been estimated to 2 TWh. This has been done by summing up the theoretical potential calculated with the elevation energy of the water in the storage reservoirs [15], [16]. In reality, the total capacity will probably be smaller because of constrains regarding minimum water height in the reservoirs, maximum outflow per day, natural inflow, rain, etc.

The capacity of the German PHES system has been set to 60 GWh with an installed pump and turbine power of 8 GW.

Rejected power and energy

The rejected energy and the maximum power surplus that appears in each scenario for Germany, Austria and the combined system are listed in Table 2. The total rejected power and energy in Germany is the highest in scenario C. This is due to a higher installed power of PV compared to the other scenarios. It can be observed that the rejected energy in the maximum RE-scenario (CC) for both countries is lower than just in scenario C for Germany. On the other side the maximum power surplus is slightly higher.

[†] On net electricity consumption, not electricity generation.

Table 2. Rejected power and energy from renewable sources in all scenarios investigated

2050 Scenario	Max. rejected power [GW]	Rejected Energy [TWh]
DE – A	51.15	21.72
DE – B	38.85	15.85
DE – C	69.09	29.04
AT – B	0	0
AT – C	12.18	7.69
BC	43.07	11.34
CC	70.54	24.96

Energy storage needs of Germany and Austria

The results of the energy storage needs calculation for Germany and Austria as isolated systems are shown in Table 3. As can be seen, there is no need for additional energy storage capacity just for the Austrian system. As there is no surplus of renewable energy in scenario AT-B the energy storage system (ESS) is operating only to smoothen the residual load and thus there is no need for any expansion. Only in scenario AT-C there is an additionally needed power in charging mode of 3 GW to fully integrate all energy from RES, see Figure 1. The charging level of the Austrian PHES system is very high and reaches almost 80% of its total capacity, which is an equivalent of around 1,600 GWh. This is due to the very high surplus of PV power in summer, which cannot be discharged during this period of the year and thus the ESS is almost continuously in charging mode. As can be seen in Figure 1, the reservoir is discharged again during autumn and beginning of winter when there is lower feed-in from PV. Another problem that could arise when trying to store all this surplus is that the reservoirs probably will not be empty at the beginning of the year and thus will not have enough capacity so store all the surplus of energy from PV.

Table 3. Additionally needed storage power and capacity for all scenarios

2050 Scenario	Additionally needed power [GW]		Additionally needed capacity [GWh]
	Charing	Discharging	
DE – A	38.79	25.17	1,308
DE – B	31.85	25.74	1,534
DE – C	55.16	29.04	950
AT – B	0	0	0
AT – C	3	0	0

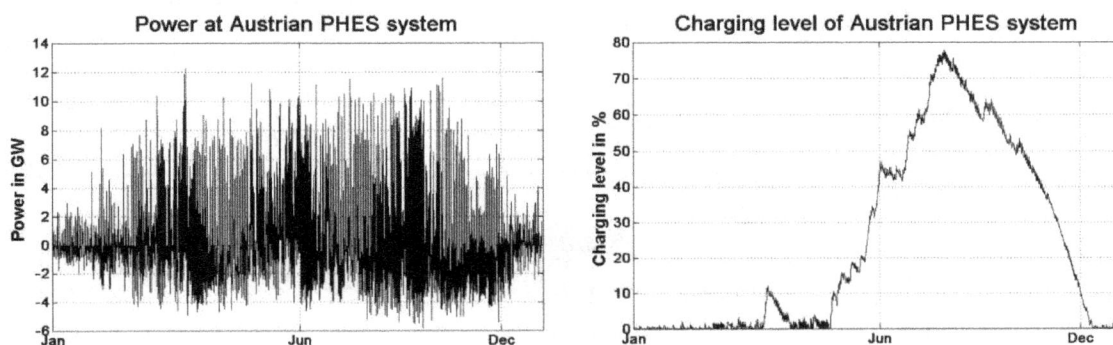

Figure 1. Used power and charging level of expanded Austrian PHES system in scenario AT-C

On the other side, when looking at the results in Table 3, there is an enormous need of new storage installations in Germany in all scenarios. The highest capacity is needed in the scenario with a favoured development of wind energy (1,534 GWh). The surplus produced by wind energy appears mainly during autumn, winter and spring. At this point the Austrian system could be used to store the surpluses of wind power during winter when there is a low feed-in from PV in the Austrian system.

For the German system the storage needs can further be divided into short and long term energy storage needs. This should be done because the German system will probably have to handle short term energy storage needs with different technologies with different efficiencies and capacities, whereas Austria will be able to rely on the PHES potentials in their own country. For the separation, maximum full load hours of 24 h, a maximum power of 12 GW in charging and discharging mode and a cycle efficiency of 85% was set as technical constraints for the short term energy storage facilities (SES), see also [11]. At this point, no constraints were set for the long term energy storage (LES). The results of this separation for Germany are shown in Tables 4 and 5. Due to the power limitations, the peaks of the RE feed-in are cut off and the overall needed power for the ESS decreases. Especially the needed power for LES is in a range of the already installed PHES facilities in Austria.

Table 4. Needed power and capacity for the long term energy storage system (LES)

Germany - LES			
	A	B	C
$P_{charging}$ [GW]	13.43	14.91	12.80
$P_{discharging}$ [GW]	6.43	6.43	7.44
Capacity [GWh]	1,106	1,206	861

When looking at the capacity factor (CF) of SES and LES in Table 5, it can be observed that the CF especially for LES is very low. The same can be observed for the Austrian ESS. When integrating all surpluses of renewable energies, the capacity factor of the PHES system is at 20.03%. Taking this into account it seems obvious to take a look at the combined system of Austria and Germany to see if there are synergies that could be used to optimise the operational strategy of both countries' energy storage systems.

Table 5. Capacity factor of short term energy storage system (SES) and long term energy storage system (LES) for all scenarios in Germany

Germany					
	A		B		C
	SES	LES	SES	LES	SES	LES
Charging	14.20%	6.65%	10.78%	5.25%	18.30%	5.95%
Discharging	14.11%	12.95%	10.68%	13.37%	18.22%	10.16%
Total	28.31%	19.60%	21.46%	18.62%	36.52%	16.11%
Rejected energy	2.91 TWh		1.72 TWh		6.64 TWh	

Energy storage needs of the combined system

The residual load of the combined system of Germany and Austria can be seen in Figure 2 for scenario CC. The strong influence of the PV can be observed in the appearance of almost daily negative peaks in the period from April to end of October. Due to low feed-in from wind in the Austrian system the surplus of wind during winter in

Germany is mostly compensated by the Austrian load demand. This can already be seen as first positive effect of the combined system. Nevertheless the negative peaks of the residual load often reach more than 50 GW, which exceeds the, up to date, installed ESS-power by far. As the transmission capacities between Austria and Germany will probably not reach 50 GW, two scenarios are investigated to show a theoretical and a realistic example. First, no limitations are set to the transmission capacities. As a second, more realistic, scenario the transmission capacity is limited to 9.2 GW – the rated power of the expected Austrian PHES system.

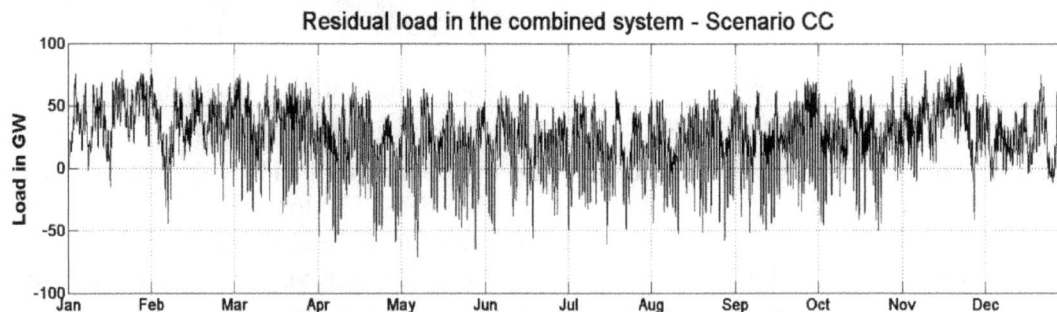

Figure 2. Residual load in the combined system of Germany and Austria, scenario CC

Without transmission capacity limitations. The used power and the charging level of the expanded Austrian PHES system are shown in Figure 3 for scenario CC. It can be observed that the degree of capacity utilization of the Austrian ESS is at 32% (640 GWh), which is less than half of the needed capacity of the single Austrian system in the GREEN-Scenario. This is due to the fact that the surplus produced during PV-summer-feed-in in Austria can be used in Germany during the night. In the single Austrian system the load during the night is not high enough and thus the charging level rises constantly during summer. The highest charging level is reached when there is a strong feed in from wind in addition to the high PV penetration. This effect can be observed in Figure 3 during spring and in August. However, the high installed amount of PV power leads consequently to a very high surplus of power during sunny days and thus very high additionally needed power especially in pumping mode, see Figure 3 and Table 6.

Table 6. Overview of the storage operation in Germany and Austria with no transmission system limitations

2050 Scenario	Country	Stored Energy [GWh]	Provided Energy [GWh]	Max. used power [GW]	
				Charge	Discharge
BC	DE	17,111.39	13,837.93	8.0	8.0
	AT	27,005.47	21,398.66	36.4	37.2
CC	DE	20,174.01	16,321.21	8.0	8.0
	AT	44,102.49	35,671.96	59.2	37.7

In scenario BC the synergy effect is less significant. The maximum charging level decreases to 39%, which is still almost half of the needed capacity of the single Austrian system. The highest charging level appears in August and is due to a stronger feed-in from wind compared to scenario CC. The additionally needed power is around 37 GW in charging and discharging mode, which is slightly lower than in the single German system.

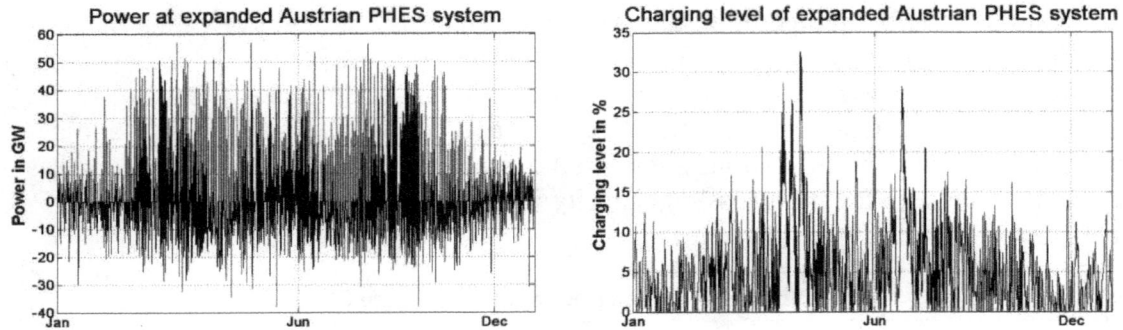

Figure 3. Used power and charging level of expanded Austrian PHES system in the combined system scenario CC

As can be seen in Table 7, the capacity factor of the German PHES system is very high in all scenarios compared to the one of the Austrian ESS. The low CF of the Austrian system is due to the fact that it has to take all the surplus of renewable energies. The maximum power of the PHES system is only used once per year which is not an economical solution. The high capacity factor of the German system is due to the daily fluctuations of PV. This ensures that the reservoir can be filled and emptied almost daily.

However, it should be kept in mind that the transmission capacity from Germany to Austria is today only 2.2 GW, which is normally already used by the existing PHES system of Austria. In this scenario transmission lines of 20 GW to 57 GW would have to be built.

Table 7. Capacity factor of the German and Austrian PHES facilities in the scenarios of the combined system

	German PHES			Austrian PHES		
	Charge	Disch.	Total	Charge	Disch.	Total
BC	24.42%	19.75%	44.17%	8.47%	6.57%	15.04%
CC	28.79%	23.29%	52.08%	8.50%	10.80%	19.30%

Power limitation of Austrian PHES system. In this scenario a more realistic approach is chosen by limiting the transmission capacity and turbine / pump power of the Austrian system to the, by 2050, expected value of 9.2 GW. The smoothened residual load after the use of the German and the Austrian PHES system is shown in Figure 4. It can be seen that there are still very high negative peaks of more than 50 GW but just for few hours during the year.

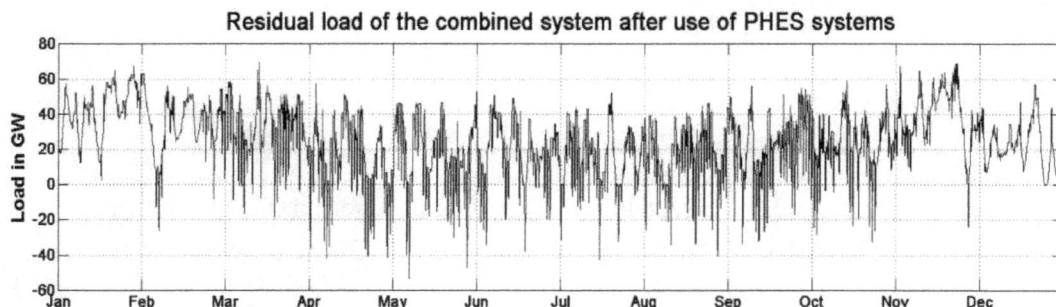

Figure 4. Residual load in scenario CC after the use of the German and Austrian PHES system without any power extensions

The rejected energy from renewables due to these limitations would be 9.02 TWh, which is less than 2% of the total possible production from RES-E. So, already with no further than the expected expansion of the PHES system and by increasing the transmission capacity from 2.2 GW to 9.2 GW, the rejected energy from renewable sources can be reduced from 24.96 TWh to 9.02 TWh. The outcome of the simulation regarding the energy storage system operation is shown in Figure 5 and Table 8. Due to the power limitations the maximal used power in charging and discharging mode is 9.2 GW for the Austrian and 8 GW for the German system.

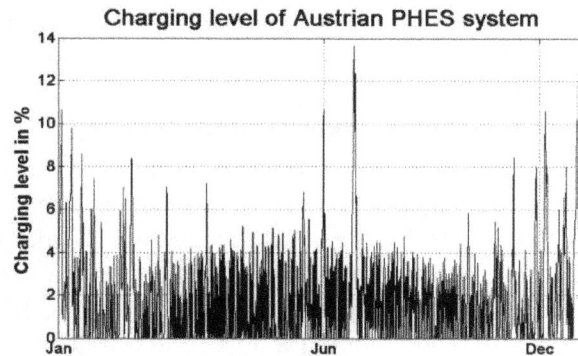

Figure 5. Charging level of expected Austrian PHES system in scenario CC with a limited transmission capacity of 9.2 GW

The most significant difference to the unlimited system can be observed in Figure 5. It can be seen that the maximum charging level of the Austrian PHES system is less than 14% of the total capacity. This is less than half of the fluctuation that appeared in the unlimited system. This could also be in a justifiable range for the big reservoirs in the Austrian Alps that are nowadays only used for seasonal energy storage with relatively low fluctuations.

Taking a look at the capacity factors of the PHES systems in both countries in Table 9, it can be observed that the CF is at more than 40% in all scenarios. For the German system this represents only a marginal difference but the CF of the Austrian system more than doubled for scenario BC and even tripled for scenario CC. This ensures that the ESS is used to a high extend and that it could be operated economically feasible

Table 8. Overview of the storage operation in Germany and Austria with a transmission capacity limited to 9.2 GW

2050 Scenario	Country	Stored Energy [GWh]	Provided Energy [GWh]	Max. used power [GW]	
				Charge	Discharge
BC	DE	20,228.96	16,365.63	8.0	8.0
	AT	20,859.28	16,762.69	9.2	9.2
CC	DE	17,161.74	13,878.71	8.0	8.0
	AT	20,326.55	16,020.09	9.2	9.2

Table 9. Capacity factor of the German and Austrian PHES facilities in the scenarios of the combined system

	German PHES			Austrian PHES		
	Charge	Disch.	Total	Charge	Disch.	Total
BC	24.49%	19.80%	44.29%	25.22%	19.88%	45.10%
CC	28.87%	23.35%	52.22%	25.88%	20.80%	46.68%

System dynamics in the different scenarios. In the scenarios investigated before, the ESS is used to smoothen the residual load of the combined electricity supply system of Germany and Austria. Figure 5 is showing some benefits energy storage can bring to the flexibility of the electricity supply system but also for the operation planning of the conventional power plant fleet. The black bars show the variation of the residual load and its occurrence over the year under investigation. The blue and the dashed red line show the variation of the residual load after the use of the particular energy storage system in the scenario CC with an unlimited and limited ESS respectively. There are three time steps in the three different subplots. The time steps are GW per 1 hour, 3 hours and 8 hours. The differentiation is made to take into account different starting times, cold as well as warm start, for different power plant types. It can be observed that without any use of ESS there is a high occurrence of high load fluctuations. Especially fluctuations of more than 50 GW/8 h appear more than 600 times and fluctuations of more than -50 GW/8 h more than 400 times per year. These high fluctuations are filtered out with the use of ESS. It can be seen in Figure 6 that especially the occurrence of fluctuations of less than 1 GW in the different time steps is increased strongly. Like this the need of redispatch measures in the daily operation can be reduced. Relevant for the investigation in this paper is the different impact of the limited and the unlimited ESS. It can be seen that unlimited system can smoothen the residual load stronger than a system with limited power. Nonetheless the smoothening is high enough to strongly increase the system stability and to enable an easier operation planning of the left over power plants. Additionally the predictability of system load variations rises as the ESS can also be used to reduce forecast errors of renewable energy production.

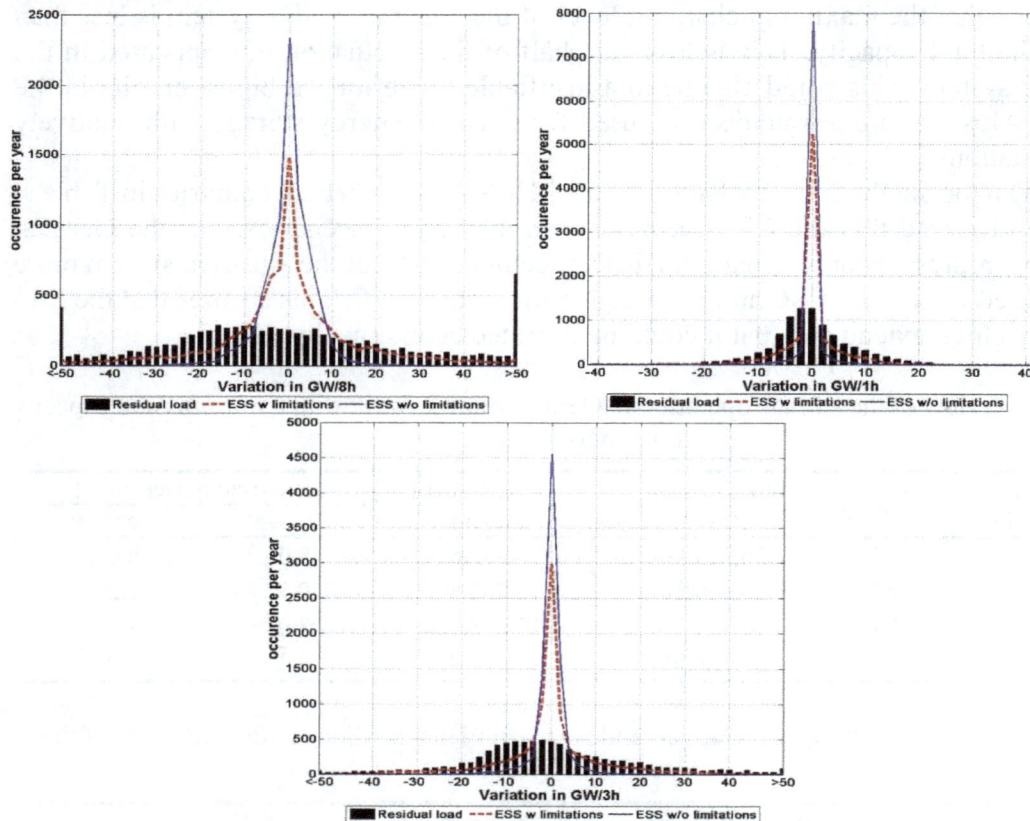

Figure 6. Load variations in scenario CC of residual load without ESS (black bars) and with the use of limited (red line) and unlimited (blue line) ESS with variations in GW per 1 hour (a); in GW per 3 hours (b); and in GW per 8 hours (c)

CONCLUSION AND OUTLOOK

It has been shown that the connection of the Austrian and the German electricity system gathers a very high potential. Especially the better exploitation of the Austrian PHES system is a big advantage for both countries. The capacity of the existing reservoirs is big enough to cover even the total storage needs of Germany. The needed transmission capacity will most probably be too high. But, even with lower capacities a very high surplus of RES-E can be stored. When building new transmission lines also the Austrian energy system becomes more flexible and the total energy storage needs in both countries decrease. All outcomes are again summarized in Table 10.

Table 10. Summary of results for all scenarios investigated within this paper

2050 Scenario	Rejected Energy		Add. needed power		Add. needed capacity	Max. fluctuation
	Before ESS	After ESS	Charging	Discharging		
DE – A	21.72 TWh	0	38.79 GW	25.17 GW	1,308 GWh	100%
DE – B	15.85 TWh	0	31.85 GW	25.74 GW	1,534 GWh	100%
DE – C	29.04 TWh	0	55.16 GW	29.04 GW	950 GWh	100%
AT – B	0	0	0	0	0	<1.5%
AT – C	7.69 TWh	0	3 GW	0	0	78%
BC	11.34 TWh	0	27.2 GW	28 GW	0	DE: 100% AT: 40%
CC	24.96 TWh	0	50 GW	28.5 GW	0	DE: 100% AT: 33%
BC w TC limitations	11.34 TWh	2.29 TWh	0	0	0	DE: 100% AT: 33%
CC w TC limitations	24.96 TWh	9.02 TWh	0	0	0	DE: 100% AT: 14%

In the stand alone German system the additionally needed power reaches from 31.85 GW to 55.16 GW in charging mode and 25.17 GW to 29.04 GW in discharging mode depending on the particular scenario. The additionally needed storage capacity reaches, depending on the scenario, from 950 GWh to 1,534 GWh. First advantages of the combined system could already be observed in Table 2. In each scenario the rejected power and energy of the combined system is lower than the sum of each value in both islanded systems. In scenario AT-B no surpluses of renewable energies occur and thus no additional capacity is needed. The ESS is used anyway to a small extend to balance out daily RL fluctuations. In scenario AT-C the high installed power of PV causes high surpluses during summer which leads to an additionally needed power in charging mode of 3 GW. Although the overall rejected energy is low compared to the capacity of the Austrian PHES system, the reservoirs are filled up to 80% (see Figure 1) due to the strong feed-in of PV during summer. The positive impact of the combined system regarding the fluctuation of the water level in the Austrian reservoirs is clearly visible, see Figures 3 and 5. In none of the scenarios of the combined system the maximum fluctuation reaches more than 40%. In scenario CC with transmission system limitations the maximum fluctuation could even be reduced to 14%. This could still be in a justifiable range compared to almost 80% in the islanded system scenario AT-C. Nonetheless without any limitations and to fully access the potential of the Austrian PHES system, new transmission capacities from Germany to Austria and vice versa in a range of 34 GW to 57 GW would have to be built. This will hardly be justifiable economically and towards the public. But, just with an expansion from nowadays 2.2 GW to 9.2 GW, only 2% of the energy from renewable sources would get lost. This percentage would be higher when using less efficient storage technologies like power-to-gas instead of building transmission lines.

However, besides the advantages of a combined system shown in this paper, there are still some investigations to be made, e.g.:

- Public opinion/opposition;
- Determining the actual usable potential in the Austrian Alps;
- Taking into account the natural inflow in the reservoirs due to rivers and rain and hence the must run moments of Austrian hydropower plants, e.g. see [2].

Furthermore, the expansion of this investigation to a European scale is planned to see what kind of influences a wider area of load consumers can have on the storage needs on the one side, and on the business models for energy storage systems on the other side. Price building concepts for the services provided by energy storage facilities will be evaluated. In this context the participation of Austrian and German PHES systems on the balancing energy markets as well as on an internal European electricity market will be investigated.

ACKNOWLEDGMENT

The results presented in this paper were mainly produced within the work on the EU founded project store (www.store-project.eu), Contract-Nr. IEE/10/222. The authors gratefully acknowledge the contribution of the partners in determining the scenarios and gathering the data needed to do this simulations.

REFERENCES

1. Nitsch, J., Gerhardt, N., Wenzel, B., et al., Long Term Scenarios for the Development of Renewable Energies in Germany, Taking into Account the Development in Europe and Worldwide, *Final Report (BMU - FKZ 03MAP146)*, in German, March 2012.
2. Boxleitner, et al., Super-4-Micro-Grid, Sustainable Energy Supply Within the Climate Change, Konsortium aus TU Wien, Zentralanstalt für Meteorologie und Geodynamik, TIWAG, Voralberger Illwerke AG, Verbund Hydro Power AG, *Final Report*, in German, Vienna, 2011.
3. Kleinmaier, et al., Energy Storage for the Energy Turnaround, ETG Task Force Energiespeicherung, *VDE Studie, Final Report*, in German, 2012.
4. Weiß, T., Overview of the Current Status and the Future Development Scenarios of the Electricity Supply System and an Estimation of Future Energy Storage needs, *Deliverable 5.1 of the Store Project*, March 2013.
5. Schulz, D., Jordan, M., Concepts for Pumped Hydro Storage Facilities in Former Open Cast Mining Locations, *5th International Renewable Energy Conference IRES 2010*, Germany, Berlin, 22-24 November 2010.
6. Weiß, T., Mattner, S., Grumm, F., Schulz, D., The Federal Waterways as Energy Storage System - Potentials and Challenges, *WasserWirtschaft 7/8 2013*, in German.
7. Eickmann, J., Drees, T., Sprey, J. D., Moser, A., Optimizing Storages for Transmission System Operation, *Energy Procedia*, Vol. 46, Elsevier, pp 13-21, 2014.

8. Schneider, M., Boras, P., Schaede, H., Quurck, L., Rinderknecht, S., Effects of Operational Strategies on Performance and Costs of Electric Energy Storage Systems, *Energy Procedia*, Vol. 46, Elsevier, pp 271-280, 2014.

9. Weiß, T., Lücken, A., Schulz, D., An Empirical approach to Calculate Short and Long Term Energy Storage needs of an Electricity System, *48th International Universities' Power Engineering Conference*, 2-5 September 2013, Dublin, Ireland.

10. Makarov, Y. V., Pengwei, D., Kintner-Meyer, M.C.W., Chunlian, J., Sizing Energy Storage to Accomodate High Penetration of Variable Energy Resources, *IEEE Transactions on Sustainable Energy*, Vol. 3, No.1, 2012.

11. Weiß, T., Schulz, D., Development of Fluctuating Renewable Energy Sources and its Influence on the Future Energy Storage needs of Selected European Countries, *4th International Youth Conference on Energy 2013*, 6-8 June 2013, Siofok, Hungary.

12. Beurskens, L.W.M., Hekkenberg, M., Renewable Energy Projections as Published in the National Renewable Energy Action Plans of the European Member States, ECN-E--10-069 http://www.ecn.nl/docs/library/report/2010/e10069.pdf, [Accessed: 16-July-2013]

13. Huber, C., Faber, T., Haas, R., Resch, G., Green, J., Olz, S., White, S., Cleijne, H., Ruijgrok, W., Morthorst, P.E., Skytte, K., Gual, M., del Rio, P., Hernandez, F., Tacsir, A., Ragwitz, M., Schleich, J., Orasch, W., Bokeman, M., Lins, C., Action plan for a dynamic RES-E policy, *Report of the European Research Project Green-X* - funded by the EC-DG Research, Vienna University of Technology, 2004.

14. Zach, K., Auer, H., Lettner, G., Weiß, T., Assessment of the Future Energy Storage Needs of Austria for Integration of Variable RES-E Generation, *Deliverable 5.1 of the Store Project*, 2013.

15. Tamerl, H., AHPs Storage Power Plants in Kärnten, Verbund-Austrian Hydropower AG, in German, 2007.

16. Tamerl, H., AHPs Hydro Power Plants in Tirol, Verbund-Austrian Hydro Power AG, in German, 2007.

10

The Project Activities of the Renewable Energy Resources Use in the Republic of Serbia

Larisa Jovanović[*], *Lazar Cvijić*
ALFA University, Belgrade, Serbia
e-mail: larisa.jovanovic@alfa.edu.rs

ABSTRACT

With the ratification of the Energy Community of Southeast Europe countries (14th July 2006) the Republic of Serbia, among other things, accepted the obligation to adopt and implement a plan of applying the Directive 2001/77/EC about promoting the production of electrical energy from renewable energy sources. The projects of the renewable energy resources use have a positive impact on the environment, in particular about the mitigation of global climate change and local environmental sustainability.

KEYWORDS

Renewable energy resources, Small hydropower plants, Landfill biogas, CDM projects, Republic of Serbia.

INTRODUCTION

The planned gross electricity production in the Republic of Serbia in 2013 was 37,649 TWh, which is 2% higher than the estimated production in 2012. The structure of the electricity production is as follows: thermal power plants: 28,150 TWh (75%), hydro power plants: 9,116 TWh (24%), others (including industrial power plant, wind and solar power plants): 1%. The main fuel is coal, primarily lignite or brown coal. The domestic production of electricity almost entirely meets the demand. Planned production of heat in 2013 was 34,401 TJ. Thermal power plants (6%), thermal-heating stations (5%), industrial power plants (25%) and heating plants (64%) participate in the structure of the production. Most of the power plants are older than 40 years and it is urgent to revitalize the existing capacities.

The Republic of Serbia started implementation of the commitments under the Kyoto Protocol [1] for the use of renewable energy sources [2]. In accordance with the National Strategy for Sustainable Development [3] and the Law on the ratification of the amendment to Annex B of the Kyoto Protocol [4], priority policy measures to mitigate the effects of climate change have been identified [5]. In order to improve the risk management system in accordance with the Plan of Spatial Planning of Serbia for the period 2010-2020 year [6], the Ministry of Infrastructure and Energy of the Republic of Serbia prepared a National strategy for the implementation of the Clean Development Mechanism (CDM) projects in the Republic of Serbia. The National strategy for the inclusion of the Republic of Serbia in CDM projects for waste disposal was adopted in February 2010. The Strategy contains information on total assets and priorities for the implementation of CDM projects in the Republic of Serbia [7]. In 2011, the first report

---

[*] Corresponding author

was adopted by the Republic of Serbia [8] for the obligations of the United Nations Framework Convention on Climate Change [8, 9].

In February 2013 the Serbian government has adopted a draft law on energy efficiency, which covers all aspects that affect the costs of municipalities. The new Energy Law clearly defines the privileged electrical and thermal energy producers with the right to subsidies and benefits. The energy producers mentioned in the Law are those which use the renewable energy sources (biogas, landfill gas, wind…) and at the same time produce electrical and thermal (cooling) energy, while fulfilling conditions with respect to energy efficiency and environmental protection.

The National Action Plan for Renewable Energy has received the endorsement of the Energy Community in accordance with the development plan of Serbia until 2020. The Republic of Serbia has committed itself to increasing the share of renewable energy in total energy consumption to 27% by 2020 [10]. 2015 will be the first reference point for judging the success of control measures in the area of increasing the use of renewable energy sources. As matters stand at the end of 2015, we should expect the trend of further development of renewable energy projects.

In the framework of the German model, local communities will be encouraged to become owners of wind generators to produce electricity. Standard contracts for the purchase of electricity from renewable energy sources will also be accepted.

Serbia expects two billion dollars of investment projects in renewable energy over the next seven years in order to introduce new green energy capacity (up to 1,092 MW). These include wind farms with the capacity of 500 MW, hydropower plants (200 MW), biomass processing plants (100 MW) and the plants using other sources of energy (192 MW), which will help stabilize the power grid, to create jobs for people in the green industry sector and prevent environmental problems.

THE CDM PROJECTS IN SERBIA

The National Authority for the implementation of projects under the Clean Development Mechanism in the Republic of Serbia was formed by the Ministry of Environment and Spatial Planning of the Republic of Serbia. It is responsible for issuing permits to work on projects under the clean development.

The main criteria for participation in CDM projects are:
- Ratification of the Kyoto Protocol;
- Participation in the project on a voluntary basis;
- Establishment of a national authority for the implementation of CDM projects under Kyoto Protocol.

The use of renewable energy sources in the energy sector of the country is an important indicator of sustainable development and energy efficiency [11-13].

To encourage development and investment in the energy sector and meet its obligations under the Energy Community of South East Europe, the Serbian government has adopted a number of regulations, including:
- Establishing a system of feed-in tariffs in which the Serbian government will subsidize the cost of renewable electricity;
- Defining the requirements for the status of privileged power producers that use renewable energy sources for electricity generation.

The system of incentives is thoroughly regulated by the Regulation on incentives for the privileged electrical energy producers, which entered into force on February 4, 2013 [14]. In particular, feed-in tariffs for electrical energy depend on the plants and installed capacity. For hydro power plants, this tariff varies between 5.9 and 12.40 c€/kWh, for biogas plants, 12.31-15.66 c€/kWh and power plants using landfill gas and gas from the plants for the treatment of municipal wastewater, 6.91 c€/kWh.

Unfortunately, the production of energy from biomass, which can be regarded as a very promising source of renewable energy in Serbia is still limited to the production of pellets from waste in agroforestry and agroindustry [15]. Serbia has significant potential for the production of raw material for processing into biodiesel, which is estimated at about 10% of the total area of arable land. This area can provide enough raw materials for the production of 210 to 250 thousand tonnes of biodiesel a year to replace about 13-16% of fossil diesel fuel in Serbia.

However, Serbia has not adjusted production of biodiesel during 2013. Its production and use are not competitive when compared with the price of mineral diesel fuel. Biodiesel is produced in Serbia in private workshops, in handicraft installations for the needs of entrepreneurs.

A key condition for the large-scale production and use of biodiesel in Serbia will be the introduction of economic measures that will enable biodiesel to be competitive compared to the prices of mineral diesel. The most important measure is the total or partial exemption from the excise tax on biodiesel. A particular problem is the unregulated market of oilseeds, which is characterized by fluctuations in the price and limitations of areas under oilseeds.

The first factory for the production of biogas from biomass in Serbia was built at the end of 2012 in Vrbas. The owner of the factory will sell electricity for 12 years to the public supplier at a price of 12.31 c€/kWh.

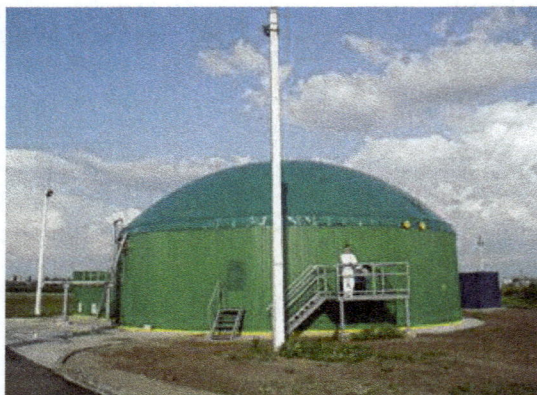

Figure 1. First in the RS factory biogas in Vrbas

BIOGAS FROM LANDFILLS

In industrial countries, between 300-400 kg of waste is produced per person annually. The waste is collected and disposed in safe and sanitary landfills which implies the protection of underground waters, and also air protection from unclean and dangerous landfill gas. The landfill gas is created by the degradation of organic materials under the influence of microorganisms in anaerobic conditions. In the middle of the landfill there is an overpressure, so landfill gas is released into environment. An average content of the landfill gas is 35-60% of methane, 37-50% of carbon dioxide, and in small quantities we can trace carbon-monoxide, nitrogen, hydrogen sulphide, fluorine, chlorine, aromatic hydrocarbons and other gases in traces. On the basis of the stated content of landfill gases, we can see that this gas is very harmful to environment, especially to living organisms and infrastructural objects in the close vicinity of landfills, because methane is very explosive under certain conditions. Methane is 21 times more harmful to climate and ozone layer than carbon dioxide, which means that 1 tonne of methane damages ozone layer (greenhouse effect) as 21 tonnes of carbon dioxide. In order to eliminate the negative effects of uncontrollable expansion of landfill gas, we perform planned collection and forced gas direction towards combustion sites. It allows faster stabilisation of fresh landfill parts, decrease of the underwater contamination, and it allows the use of energy produced in a landfill (heating, hot water, electricity, etc.).

The legal obligation of collecting and flaring landfill gas will be the right solution: combustion gas for energy purposes while creating economic benefits [16].

Benefits from the air pollutants reduction make a valuable contribution to sustainable development [17].

This concept implies placing vertical perforated pipes in the body of the landfill (wells, probes) and their horizontal connection. Through a compressor plant the landfill gas is sucked, compressed, dried and directed towards the gas motor. For security reasons, the installation of a high-temperature torch is recommended, because it flares the excess gas.

The landfill gas with an average content of methane of 50% has a lower thermal value in the range of 5 kWh/m^3, which makes it good fuel for gas engines specially designed for this purpose.

The gas engine drives a generator to produce the more valuable electricity. Through the heat exchanger, we obtain the heat energy from the water that cools the engine and lubricating oil, and also from the exhaust gases. It is possible to attain a high degree of beneficial effects with the combination of electrical and thermal energy (electrical 40%, thermal 43%). This means that, from 1 m^3 of landfill gas, we get 2 kWh of electrical energy and 2.15 kWh of thermal energy. The produced electrical energy is used for our own purposes, or it is directed into the electrical network. The produced heat is used at landfills for the production of hot water, or in conservatories where early vegetables and flowers are produced. It is also used in industrial plants, in the vicinity of landfills, or for heating of apartments. Implementation of the CDM in the dumps in the cities of Uzice, Cacak and Pozega, includes the preparation of three landfills for closure procedure.

The added value of the project is to enable the collection and combustion of biogas from landfills. Biogas is a complex mixture of gases formed from organic waste as a result of microbial activity. Maintenance work on the project requires a capital investment in gas gathering system and the purchase of equipment for the combustion of methane. The project will reduce the risk of pollution for residents of areas in the immediate vicinity of the existing landfills. Landfills are located in the municipalities of central Serbia (Uzice, Pozega, Cacak). These municipalities, along with six other municipalities in the area of Zlatibor agreed to establish a new landfill. Hence, these existing landfills will be out of action for the next three years.

This project activity would involve investments in gas collection systems and equipment for methane flaring. The project will also reduce environmental risks for the residents in the vicinity of the existing disposal sites.

It is estimated that the project can reduce emissions through flaring of captured landfill gas of nearly 140,000 t CO_{2eq} in total for the period 2008-2018, or given split to each dump site 52,000 t CO_{2eq} in Uzice, 24,000 t CO_{2eq} in Pozega and 64,000 t CO_{2eq} in Cacak.

In recent times, there has been a greater use of biogas extracted from landfills and wastewater. Even if it is not used for thermal and electrical energy, the landfill gas must be properly disposed and cleaned, because it contains harmful combustible elements, many of which form smog. Biogas digesters use biodegradable materials, of which two useful products are obtained: biogas and high quality fermented bio fertilizer. Biogas filtered to the level of purity for pipelines is called renewable natural gas, and it is possible to use it on every occasion in which natural gas is used. This includes transport of gas through pipelines, electricity production, space and water heating, and in many technological processes. Compressed biogas can be used to power motor vehicles.

The basic production process of biogas mainly consists of three parts:

- Preparation of biological input;
- Degradation;
- Waste treatment.

Biogas (and biomass) is renewable energy source with low amounts of carbon. If used properly, biomass is sustainable fuel that can considerably influence the decrease of carbon net emissions compared with other fossil fuels.

Although burning of biogas and natural gas leads to the creation of certain amounts of carbon-dioxide (CO_2), carbon in the biomass originates from plant mass, which contains carbon from atmospheric carbon dioxide. Thus, the use of biogas is viewed as CO_2 neutral and does not influence the greenhouse effect with the increased amounts of gases.

Further conclusion can be drawn from the fact that the substitute of fossil fuels for biogases brings a decrease of CO_2 emissions.

Carbon, which is present in biomass of approximately 50% of its dry matter, is already a part of atmospheric cycling of carbon. Plants absorb CO_2 from the atmosphere during its entire life cycle. At the end of the plant life cycle, the same carbon returns to the atmosphere as a mixture of carbon dioxide and methane. Methane is transformed into carbon dioxide in the atmosphere, and in this way the cycle is completed. The burning of biomass and biogases, whether direct or indirect, as a fuel, also releases CO_2 into the atmosphere. Of course, this CO_2 is part of the cycle of carbon in the atmosphere.

The international waste management company A.S.A. had built the landfills for communal waste, all completely in accordance with the European standards. Landfill in Lapovo was put in operation in 2009. It is currently in the procedure for approving the regional plan for waste management, after which landfill will officially acquire the status of a regional landfill, pursuant to current laws. It was built to accept all waste coming from that region, so that it may also be put at the disposal of some other municipalities, such as Novi Becej, Zitiste and others. The project development will run parallel to the implementation of investments because this project is being implemented in phases. According to the plan, what follows is the construction of new cassettes for the landfill and waste management centre, which will be equipped with new waste treatment lines.

Table 1. The CDM Projects of biogas plants in Serbia [18]

Proposer of the project	Project title
Biogas power plant Alibunar	Biogas power plant Alibunar
Nucleus Center Ltd.	Biogas Plant Raca
Bionersis SA	Collection of landfill gas at landfill Valjevo
Camco Carbon International Ltd.	Alibunar biogas plant construction project
Amest SRL	LFG recovery and electricity production at the Bubanj landfill site, Nis, Serbia
"Lazar" Ltd.	Dairy "Lazar" Biogas Project
Alltech Fermin	Energy utilization of methane obtained from improved wastewater treatment in Alltech Fermin, Senta, Serbia
PKB "IMES" Comp.	"Pigs for kilowatts" - management of animal waste and obtaining energy, Belgrade, Serbia

The landfill Vrbak in Lapovo spans nearly 22 hectares. A.S.A. Vrbak renders services of communal waste collecting and disposal for the municipalities of Lapovo, Batocina, Despotovac and Raca, whereas Topola only disposes of its waste at this landfill. A.S.A. Eko in Serbia offers services of waste collecting, treatment and disposal to citizens and municipalities, and there is also a special set of services such as maintenance of green areas, road management and cleaning, winter and summer maintenance of roads, etc. Among the users are companies, small entrepreneurs and market chains that either outsource or directly use the services of waste management, which is, in such chains, primarily related to secondary raw materials.

THE PROJECTS OF SMALL HYDROPOWER PLANTS IN SERBIA

Stocks of renewable energy resources in Serbia are high. For example, the number of favourable sites on the rivers for the construction of small hydropower plants, according to experts, is more than eight hundred.

However, permission to build small hydropower plants has been given to about 60 projects. Very often, the work on the project is suspended because of difficulties in obtaining permission based on the assessment of environmental impact. In addition, a significant problem is the financing of the construction and purchase of equipment. Not all projects are covered with matching funds [19, 20].

In cooperation with 17 municipalities in Serbia, the Ministry of Energy has prepared a tender for the release of 317 seats for the construction of mini hydropower plants in Serbia.

According to data from the Cadastre of small hydropower plants on the territory of Serbia without autonomous province Vojvodina, Serbia has the potential to build 856 small hydropower plants (100 kW to 10 MW) [21]. The total hydropower potential, gross potential water which flow through watercourses on the territory of the Republic of Serbia amounted to 25,000 GWh/year. The largest part of the hydropower potential of over 70% is concentrated in only a few watercourses with a potential of over 1,000 GWh/year: the Danube, Drina, Velika Morava, Lim and Ibar. The other rivers' potential will only be able to partially take advantage of the priority of the water management of water use, as because some rivers are planned as a source of regional water systems: Toplica, Crni Timok, Rasina, Studenica, Veliki Rzav, Mlava, Lepenac, etc.

Table 2. The Projects of small hydropower plants in Serbia [18]

Proposer of the project	Project title
IGIS IMAKO Ltd.	Construction of small hydropower plant "Gornjak"
IGIS IMAKO Ltd.	Construction of small hydropower plant "Metovnica"
IGIS IMAKO Ltd.	Construction of small hydropower plant "Ribare"
IGIS IMAKO Ltd.	Construction of small hydropower plant "Zlot"
IGIS IMAKO Ltd.	Construction of small hydropower plant "Prevalje"
IGIS IMAKO Ltd.	Construction of small hydropower plant "Jelašnica"
IGIS IMAKO Ltd.	Construction of small hydropower plant "Drvnik"
IGIS IMAKO Ltd.	Construction of small hydropower plant "Mezdreja"
EcoEnergo Group Ltd.	Construction of small hydropower plant "Jabukovik" on the City River in Crna Trava
HIDRO ONE Ltd.	Construction of a system of small hydropower plants on Ljutina river

Technically exploitable potential in Serbia is around 19.8 TWh/year, of which 17.5 TWh/year in facilities larger than 10 MW. The total technical potential of hydro power plants up to 10 MW is estimated to be about 1,800 GWh per year. The remaining technical hydropower potential and the possibility of its utilization will be determined in accordance with the criteria of non-energy related to multipurpose use of water and on the basis of political agreement on the sharing of hydropower potential with neighboring countries. Today work is being done on a detailed audit of location, to produce a more manageable list of the locations and create a better planning basis for the use of renewable sources.

This year, 2014, Serbia is beginning the construction of small hydropower plants on waterways in 156 locations.

CONCLUSION

In the Republic of Serbia, there is a need for the implementation of projects that help the environment protection. One of the goals of environmental protection is the reduction of greenhouse gas emissions according to the Kyoto Protocol. This goal can be reached, among others, by the implementation of CDM projects.

The success of these projects is crucial for further implementation of the Environmental Strategy of the Republic of Serbia, not only at national level, but also at the local level throughout the Republic of Serbia. The CDM projects have a positive impact on the environment, in particular about the mitigation of global climate change and local environmental sustainability. The CDM (Clean Development Mechanism) project activity in the Republic of Serbia involves bundling of three dump sites in order to give added value to the envisaged method of the environmentally sound closure of the dump sites. Currently, the closing procedure envisages only covering with a soil layer, leaving the possibility to consider other measures depending on waste and landfill characteristics. The added value of the project refers to enabling LFG (Landfill gas) collection and flaring at the sites. Through flaring, the methane is converted into CO_2 and hence, the greenhouse gas effect is reduced. This project activity would involve investments in gas collection systems and equipment for methane flaring. The project will also reduce environmental risks for the residents in the vicinity of the existing disposal sites.

Bundled dump sites are situated in Central Serbia, of which Uzice and Pozega in Zlatibor District, and Cacak in the Morava District. These municipalities together with six other municipalities in the Zlatibor District agreed to found a new regional sanitary landfill and the existing dump sites should be put out of operation within next three years.

The implementation of CDM projects contribute to the use of other renewable energy sources: wind power (wind turbines, wind farms), water energy (small hydro plants), solar energy (solar panels, photovoltaic systems), synthesis of biofuels from biomass, etc.

In accordance with the resolution, this will contribute to an increasing interest in the use of solar panels on the roofs of buildings. The duration of stimulating measures is 12 years.

Since the production of electricity from renewable resources in most cases is more expensive than energy production from fossil fuels, the so-called support systems have been introduced, i.e. financial and nonfinancial measures of incentives to invest in facilities that use renewable energy.

The intensive exploitation of the renewable resources must become one of the important goals of economic development in Serbia.

ACKNOWLEDGEMENTS

This study is part of the Interdisciplinary research projects III 47009 and III 43009 which are supported by the Ministry of Education, Science and Technological Development of Serbia in the period 2011-2014.

REFERENCES

1. Kyoto Protocol to the United Nations Framework Convention on Climate Change, United Nations, 1998.
2. Report on the Progress in the Implementation of the National Sustainable Development Strategy of Serbia for 2010, the Government of the Republic of Serbia. Agency for Environmental Protection, September 2011 (in Serbian)
3. National Strategy of Sustainable Development for the period 2009-2017, *Official Gazette of the Republic of Serbia*, No 57/08 (in Serbian)
4. Law on Ratification of the Amendment to Annex B of the Kyoto Protocol to the UN Framework Convention on Climate Change, *Official Gazette of the Republic of Serbia*, No 36/09 (in Serbian)
5. Law on Environmental Protection, *Official Gazette of the Republic of Serbia*, No 135/04 and 36/09 (in Serbian)
6. Law on Spatial Plan of the Republic of Serbia in Period 2010-2020, *Official Gazette of the Republic of Serbia*, No 88/10 (in Serbian)
7. National Strategy for the inclusion of Serbia into Clean Development Mechanism, for waste management, agriculture and forestry, *Official Gazette of the Republic of Serbia*, No 30/10 (in Serbian)
8. First National Report of the Republic of Serbia under the Framework Convention on Climate Change, Government of RS, 2011 (in Serbian)
9. United Nations Framework Convention on Climate Change, United Nations, 1992.
10. Gazette of Electrical Network of Serbia, EMS, Vol. 62, 2013 (in Serbian)
11. Jovanović, L. and Tomić, A., The implementation of renewable energy sources as a condition of energy efficiency in Serbia, *Ecologica*, Vol. 62, pp. 234-239, 2011 (in Serbian)
12. Jovanović, L., Tomić, A., Cvijić L., Implementation and promotion of the Sustainable development strategy of the Republic of Serbia till 2020 year, *Ecologica*, Vol. 66, pp.173-177, 2012 (in Serbian)
13. Munitlak-Ivanović, O. and Golušin, M., Concept, characteristics and condition of sustainable development indicators in selected countries of South East Europe, *Ecologica*, Vol. 63, pp. 355-362, 2011 (in Serbian)
14. Regulation on incentives for the privileged electrical energy producers, *Official Gazette of the Republic of Serbia*, No 8/13 (in Serbian)
15. http://www.energetskiportal.rs [Accessed: 06-Sep-2013]
16. Donevska, K., Jovanovski, J., Jovanovski, M., Pelivanoski, P., Analyses of Environmental Impacts of Non Hazardous Regional Landfills in Macedonia, *Journal of Sustainable Development of Energy, Water and Environment Systems*, Vol. 1, No. 4, pp. 281-290, 2013.
17. Eto, R., Murata, A., Uchiyama, Y., Okajima, K., Assessment of Co-benefits of Clean Development Projects Based on the Project Design Documents of India's Power's Sector Currently under Registration and Validation, *Journal of Sustainable Development of Energy, Water and Environment Systems*, Vol. 1, No. 4, pp. 326-339, 2013.

18. National Authority for the implementation of the Clean Development Mechanism of the Kyoto Protocol (DNA), http://80.93.243.155/DNA/?lang=src [Accessed: 26-Feb-2013] (in Serbian)
19. Golušin, M., Dodić, S., Vučurović, D., Ostojić, A., Jovanović, L., Exploitation of Biogas Power Plant - CDM Project, Vizelj, Serbia, *Journal of Renewable and Sustainable Energy,* Vol. 3, No. 5, 052701, pp. 1-11, 2011.

20. Golušin, M., Dodić, S., Vučurović, D., Jovanović, L., Munitlak-Ivanović, O., Sustainable Energy Management in Industry of the Republic of Serbia, Biogas Power Plants Advantages, *Industrija,* Vol. 40, No. 4, pp. 107-124, 2012.
21. Land Registry of small hydro power plants in Serbia, Energoproject-Hydroengineering, 1991, Belgrade.

11

Optimal Renewable Energy Systems for Regions

Karl-Heinz Kettl, Nora Niemetz, Michael Eder, Michael Narodoslawsky[*]
Institute for Process and Particle Engineering
Graz University of Technology, Graz, Austria
e-mail: narodoslawsky@tugraz.at

ABSTRACT

Most sources for renewable energy can be deduced from solar radiation as the main natural income of society. Contrary to conventional fossil and radioactive energy resources that are mined or pumped out from central point sources, solar energy is a de-central resource that requires area for its conversion to useful products and services. This requires a new technological as well as logistical concept for energy systems where regions play a key role as providers of energy and goods. The contribution will provide the conceptual framework for renewable energy system generation on a regional level, taking into account the responsibility of regions to provide goods and services to the larger society and to support urban centres. It will show how optimal resource-technology-demand networks may be constructed, using process network synthesis approaches and how the ecological efficiency of such regional systems can be measured. Application of these methods to real life case studies (in particular the region of Mühlviertel in Austria) will on the one hand prove the versatility of the methods presented and on the other hand will provide insight into the scope of necessary change if society moves towards a low carbon sustainable energy system.

KEYWORDS

Process Network Synthesis, RES, regions, Sustainable Process Index

INTRODUCTION

There is general agreement that fossil resources are approaching their production maximum. The time frame ranges up to 2020 for crude oil and up to 2060 for natural gas [1, 2], with coal remaining available for considerably longer time spans. These resource limitations have to be seen in combination with the discourse about global warming that requires a drastic reduction of (fossil) carbon emission. Taken together these two trends call for a dramatic change in the resource base over the 21st century, away from fossil towards renewable sources.

The change towards renewable resources however entails an equally drastic transformation of supply chains: whereas fossil resources are retrieved from typical point sources, most renewable resources are based on solar radiation either directly (photovoltaic) or indirectly (wind power, hydro power, biogenic resources based technologies) and therefore require area for their generation [3]. This puts new responsibilities into the hands of societal and political entities that exert control over land, most notably regions.

Interestingly enough, regions have become dynamic political players, most notably since the Earth Summit in Rio de Janeiro in 1992 [4, 5]. As the flip side of globalisation

---

[*] Corresponding author

regional and local entities have emerged as major drivers of political change in Europe [6]. These entities however have encompassing planning objectives that not only address the purely technological side of resource utilisation but also have to bring environmental and social aspects of innovations in line with economic considerations [7] and have to address issues of spatial planning, energy provision and use [8].

It is within this framework that innovation for sustainable regional energy systems has to be discussed. This requires a comprehensive set of planning tools that will be discussed and that will be elucidated in the case study offered in in this paper.

FRAMING THE PROBLEM

Providing solutions that allow regions to address their future role as major players in the game to provide society with energy and material resources requires a comprehensive approach to resource utilisation that also takes into account the inherent mechanisms of regional decision making. Renewable energy systems are characterised by highly complex interaction between actors from different sectors as well as long ranging decisions about the economic and social structure of regions and its impact on nature. Therefore decisions on the technological solutions to utilise regional resources have to be subjected to participatory planning processes involving all parties contributing and concerned by the final outcome. These planning processes by definition involve not only experts in the energy field but also providers of resources (e.g. farmers), grid operators, regional authorities and the citizens in the region that might be affected by changes in land use and energy provision as well as energy utilisation patterns. Rather than providing fixed technical solutions participatory planning requires the provision of sound, comprehensive and comparable scenarios that form the base of a discourse about the future of the region.

From a more technical point of view this requires to provide regional decision makers with the means to generate systemic structures for utilising regional resources optimally within the framework of available sources, existing economic and technical structure and demand in the region. Any planning approach that just builds on optimising single lines of resource utilisation (say optimising the use of wood) or focussing on single technologies (say biogas generation) will be insufficient to meet the planning goal of optimal resource utilisation in a region. Regions usually offer a variety of renewable resources and require meeting different demands like residential and industrial heat/cooling, electricity and mobility. This alone requires a technology system rather than optimising single technologies or the utilisation of single resources. On top of that efficiency in resource utilisation calls for interaction of technologies, where cascades of utilisation will offer higher value added on the same (limited) resource base.

Equally suboptimal are planning approaches looking for just meeting the demand within a region. Most energy forms (with the notable exception of thermal energy) are transportable and inter-regional distribution grids as well as transport pathways for concentrated energy carriers, gas and electricity exist in most regions. This subjects these energy forms to inter-regional and in many cases global market forces. It is within this inter-regional and global playing field that decision makers have to shape the future of their regions.

The task at hand for planning of renewable energy systems for regions is therefore to generate scenarios for utilisation networks that link resources, technologies, regional demand and inter-regional markets in a way that optimises the value generated for the region. This value however is not restricted to the economic aspect but also includes environmental sustainability as well as social and cultural aspects. Changing the boundary conditions of this optimisation like different land use regimes, different price structures for resources, products and services as well as taking into account competition between different uses of resources (e.g. between food and energy generation) will then lead to the decision support system needed in shaping future development in regions.

Looking at this problem from an engineering perspective, there are some aspects that can be supported by existing methods especially used in process engineering. The generation of regional technology networks is similar to the generation of optimal process networks, solved by process synthesis approaches. Both aim at generating a network of process steps that convert material and energy resources into valuable products where both resources and product demand may be limited and where different chains of process step may compete for the same resources, leading to similar products. Providing insight into the ecological pressure of regional technology networks is, at least on the metabolic level that takes into account mass and energy exchange with the environment, similar to the problem of environmental evaluation of industrial processes. It is therefore sensible to adapt the methods already well developed for process industry to the new task of providing decision support systems for regional renewable energy systems. It has to be reiterated at this point however that the results generated by these methods aim at providing scenarios for regional participatory planning rather than "optimal solutions" as they usually do in process industry.

ADAPTING PROCESS SYNTHESIS AND ECOLOGICAL PROCESS EVALUATION METHODS

There exist a wide variety of process synthesis and ecological process evaluation methods that can be adapted to the requirements of supporting planning for regional renewable energy systems. The current paper will discuss two particular methods and their adaptation and apply them to a case study.

Process network synthesis (PNS) using the P-graph method

The PNS method [9] has been successfully applied to develop optimal process networks for renewable resource utilisation processes [10-12]. This method derives maximum structures (encompassing all feasible structures fulfilling the given boundary conditions) via combinatorial rules using the bipartite graph representation of processes, arriving at optimal structures (that optimise a given target function e.g. value added generated by the process network) using a branch-and-bound optimisation routine. Besides short computation times this method has the advantage to securely find the optimum structure even for complex problems. This advantage is important in the application to regional renewable energy systems as it guarantees that all developed scenarios are actually optimal within their boundary conditions and therefore directly comparable. The method requires knowledge about the energy and material balance as well as economic parameters like operating costs, investment costs and depreciation periods for each technology included into the considerations.

A comprehensive description of the method is out of scope of the current paper; the reader is kindly referred to the original literature as well as to the very informative web-page of the PNS method [13]. The following paragraphs will be dedicated to the explanation of necessary changes and amendments to the method, if it is to be applied to regional renewable energy systems.

The main challenge by applying the PNS to regional energy systems lies in defining new "technologies" that play major roles in any resource-technology-demand network. This is in particular true for all activities within the primary sector like agriculture and forestry. Here we have one basic resource which is land. This resource is then the "input" to competing "primary technologies" i.e. different ways of land use which generate the material resources then utilised in energy technologies, be they crops, wood, grass also including residues like straw. Restrictions on land use to be considered are on the one hand climatic: not all crops may be grown in all regions. This is best handled by providing a regionally adapted set of primary technologies that generate the agricultural and forestry products amenable to the individual regional context. All these primary technologies have to be described in terms of their material and energy input (e.g. fertiliser and machinery use per hectare for a certain crop, yields per hectare and year) and their cost factors (cost of fertiliser, investment for farm equipment, etc.). Different agricultural practices (e.g. conventional and organic farming) can easily be integrated by changing the material and energy inventory as well as the prices of crops accordingly, leading not only to scenarios describing the most optimal land use but even giving decision support for the way the land is actually managed.

On the other hand there are restrictions regarding the land use as such as fields, grass land and forests are not interchangeable in regions without limitations and maintaining fertility in many cases requires crop rotation. This can be handled in partitioning the basic resource land into sub-resources such as fields, forests and grass land, each serving a particular set of primary technologies that generate the respective products, wood for forests, crops for fields and grass for grassland. Partitioning even further can be used to include crop rotation. If for instance oil seeds may only be grown every fourth year, it means that a fourth of the field area is open as a resource for the primary technology of growing oil seeds whereas the other land is not defined as a resource for this primary technology.

Finally energy technologies compete for products from primary technologies with other uses, most importantly the food sector. Therefore these products will also be assigned prices and a set of secondary technologies (e.g. husbandry, food processing) has to be included to decide between different pathways for utilising bio-resources. In many cases these technologies may also provide input to energy technologies (e.g. manure that may be used in biogas fermenters) further interlinking the maximum structure for regional applications of the PNS.

Besides including the primary sector regional renewable energy systems are critically dependent on logistics. Many biogenic resources, especially residues (e.g. straw) and wastes (manure) have dismal logistical properties like low transport densities and high water content. This means that transport is a major factor in the design of regional technology networks and has to be factored into the decision about the optimal sizes of energy provision technologies. This may be accomplished by implementing transport as intermediate technologies between biogenic resources (as products from primary technologies and/or technologies from the food sector) and different sizes of energy technologies: smaller size technologies may then be served by (local) tractor transport over a mean distance defined by regional context, installations with larger capacities require transport via road or rail according to the mean distance to their resource base, which again is dependent on regional context.

Providing heat (or cooling) for industry and residential areas is always a major factor of regional energy systems that has to be integrated into any synthesis of technology networks. This factor has two aspects: on the one hand energy provision here competes with energy saving measures and on the other hand thermal energy may only be transported over short distances via heat/cooling distribution grids. The former may be tackled by introducing "efficiency technologies" like insulating buildings. These technologies "provide" the energy difference between the situation in status quo and a situation when the optimised technology network is implemented. Investment cost, operating cost (if applicable) and material balance for these technologies, as well as energy saving per unit of technology (e.g. kilogram of insulation) have to be defined. The latter may be given for different applications (e.g. buildings of different standards).

The particular logistic property of thermal energy that it can only be feasibly transported over relatively short distances by heat/cooling distribution grids has to be factored in by indicating the heat/cooling load that might be covered by district heating/cooling. This thermal load may then be supplied either by central heating/cooling installations or by off-heat from Combined Heat and Power (CHP) plants or by excess heat from industrial plants. Conversely high temperature process heat may either be provided directly or as excess heat from CHP plants.

Another important feature of the PNS method is the possibility to balance production with demand. This is particularly useful for implementing boundary conditions often asked for by regional actors: to guarantee supply of certain goods (e.g. food) or services (e.g. residential heating) from local resources.

The Sustainable Process Index (SPI)

This index describes the aggregated ecological pressure of a certain process by the area needed to embed this process sustainably into the ecosphere, rendering a kind of "ecological footprint". The SPI identifies the area A_{tot} necessary to embed a life cycle providing a certain goods or service sustainably into the ecosphere. The life cycle comprises all activities from raw material generation to the final conversion and, when applicable, end use of a product. A_{tot} is calculated according to

$$A_{tot} = A_R + A_E + A_I + A_S + A_P \qquad (1)$$

The areas on the right hand side are called "partial areas" and refer to impacts of different productive aspects. A_R, the area required for the production of raw materials. A_E is the area necessary to provide energy. A_I, the area to provide the installation for the process, A_S is the area required for the staff and A_P is the area for sustainable dissipation of products and by-products. The reference period for these partial areas is one year. All material flows and energy flows exchanged between the life cycle to provide a good or service in question and the environment will give raise to an according area under the categories identified above. The SPI method is based on the comparison of natural flows with the flows generated by a technological process. The conversion of mass and energy flows into area is based on two general "sustainability principles":

Principle 1: Anthropogenic mass flows must not alter global material cycles; as in most global cycles (like the carbon cycle) the flow to long term storage compartments is the rate defining step of these dynamic global systems, flows induced by human activities must be scaled against these flows to long term stores.

Principle 2: Anthropogenic mass flows must not alter the quality of local environmental compartments; here the SPI method defines maximum allowable flows to the environment based on the natural (existing) qualities of the compartments and their replenishment rate per unit of area.

Whenever a life cycle produces more than one product or service (e.g. in CHP technologies where heat, electricity and material products like manure from biogas plants or ash from incineration are produced) ecological pressures have to be allocated to them according to an allocation rule. In this case study ecological pressures were allocated to all products produced in the region. Allocation was based on the income calculated at market prices.

The SPI already draws on an extensive data base concerning energy and efficiency technologies that is accessible on the web page [13] or from previous work [14, 15]. A particular tool for evaluating the impact of primary sector technologies was recently developed and is accessible via [16].

The advantage of using the SPI method for evaluating regional renewable energy systems is twofold: on the one hand this measure offers a comprehensive, life cycle wide evaluation that rates very distinct impact like CO_2 and heavy metals emissions on an aggregate level, allowing for comparison on the base of sound sustainability principles. On the other hand the SPI clearly distinguishes between renewable and fossil resource based technologies which is of high importance to regional actors.

CASE STUDY MÜHLVIERTEL

The case study will provide insight into the application of the methods described above in a real world development process on the regional level in Austria. The task at hand was to provide regional decision makers with a reliable base for deciding about the future pathway to utilise their renewable resources and restructure their energy system in order to reduce the overall ecological pressure.

The region in question is the Mühlviertel, a region spanning from the Danube to the German and Czech boarder, close to Linz, the capital of the federal state Upper Austria. The region encompasses 3,080 km² with a population of approx. 268,000 citizens. It is a highly agricultural region with particularly strong emphasis on grass land and forestry.

In co-operation with regional actors three main scenarios were defined:

- Optimal scenario: maximum value added for the region;
- Autarky scenario: total autarky for food and energy;
- Supply Linz scenario: optimal value added with responsibility to keep supply of food for the urban centre of Linz slightly above current levels.

Based on the climatic situation of the Mühlviertel and in consultation with local experts a list of possible agricultural products, their yields and limitations was defined. The current status and number of buildings as well as information about existing energy installations, waste flows and industrial energy demand was collected from a survey among all involved communities. In consultation with decision makers in the region the list of eligible technologies was defined, using a conservative approach by including only technologies that are either already state of the art or proven in industrial size demonstration plants such as the "Green Biorefinery", a technology that uses pressed juice from silage to obtain amino acids and lactic acid [17]. Together, all render a maximum structure employing the PNS method as given in Figure 1.

By setting the demand according to the boundary conditions of the scenarios and using market prices for all products and services (if not stated otherwise in the explanation of the scenarios below), the three scenarios were then calculated using the PNS editor from the homepage given above. The definition of all boundary conditions and technology parameters is however out of scope for the current paper. The interested reader is kindly referred to the end report of the project [18]. The following paragraphs will be dedicated to describe the results of these calculations as well as the ecological implications revealed by the evaluation with the SPI.

Optimal scenario 1

The boundary conditions for the optimal scenario resulted in two almost equally attractive structures for the regional technology system:

- A "biogas fuel" scenario (scenario 1A);
- "High price beef" scenario (scenario 1B).

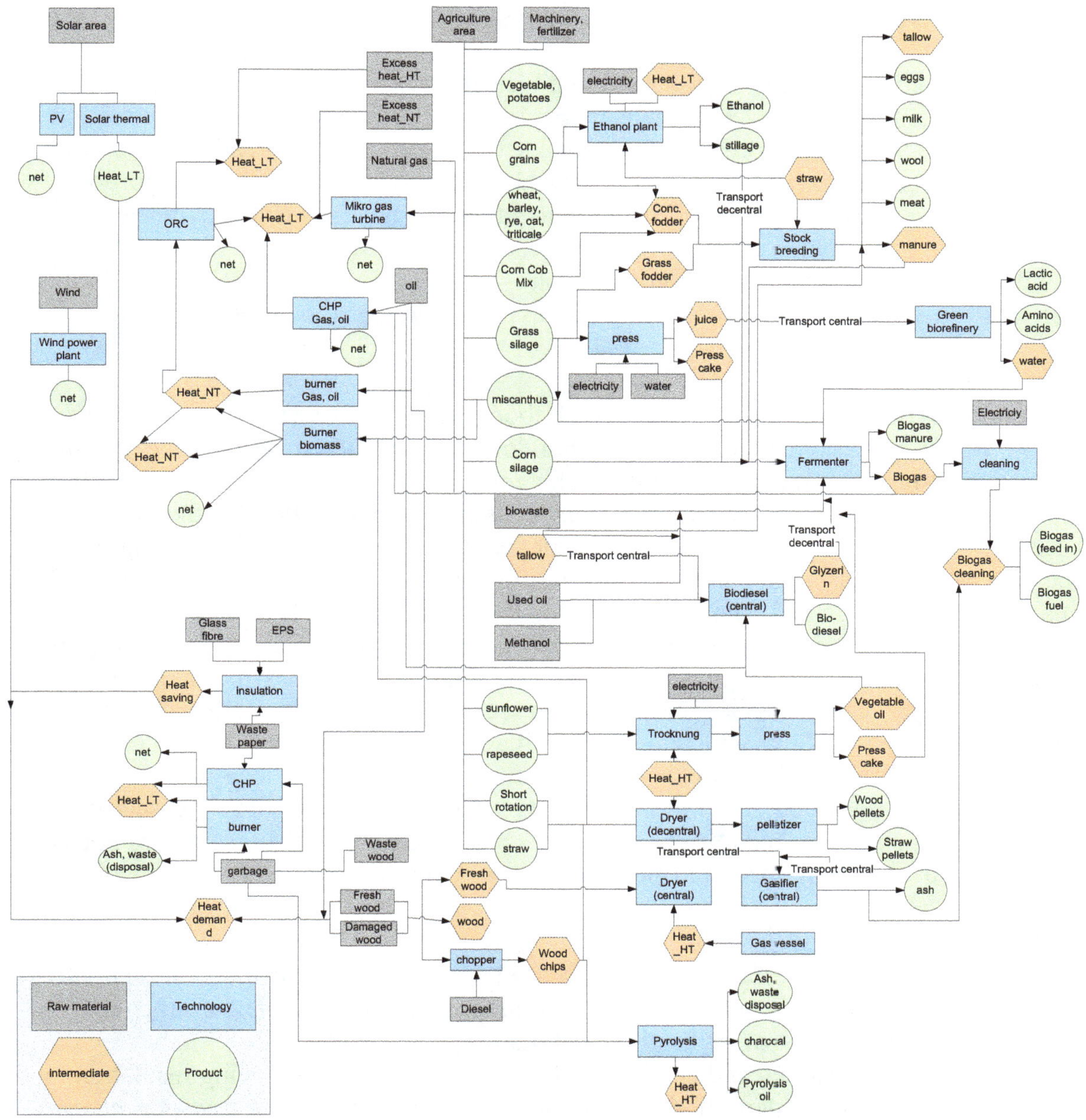

Figure 1. Maximum structure for the Mühlviertel case study

If the price for biogas is set to a level currently paid for as fuel (65 €/MWh), almost the whole grassland is used to provide input for biogas fermenters. Silage is produced from grass, then pressed, with the juice going to the green bio-refinery and the press cake is utilised in biogas fermenters. The biogas is then cleaned and fed in the grid to be distributed to fuel stations within and outside the region. Fields are mainly used to support (organic) pork breeding, with most of the pork being exported out of the region. Vegetables for regional consumption are also grown on the fields.

Almost as much added value can be achieved for the region if beef is produced with organic farming and the price for this product will be in the upper range for high quality meat (4,030 €/t). In this case grassland will be used to support cattle breeding. Manure is collected as much as possible and processed in biogas fermenters, again cleaned and fed to the grid. This scenario however has lower biogas production and no production of chemicals from the Green Biorefinery. Fields support mostly cattle breeding, with the remainder going to organic pork breeding and vegetable production for regional consumption.

Both scenarios use the available forest products for provision of residential heating as well as process heat. Wherever possible district heating based on wood chips is preferred, with firewood furnaces supplying houses outside the range of district heating distribution grids. Buildings are insulated as much as possible. All waste wood and a small portion of fresh wood are utilised in pyrolysis plants, generating oil that is subsequently refined into bio-fuel. Fat from slaughterhouses is also processed to bio-diesel in both scenarios. Wind-power, hydro power and photovoltaic are utilised to capacity which means in the case of PV a steep increase of installed area, up to 30 fold the amount used currently.

Autarky scenario 2

Autarky requires a different strategy as all food and energy have to be produced in the region. In this scenario heat for individual buildings outside the range of district heating grids is again provided mostly by wood, but here material utilisation of wood for regional construction is competing for this resource. Grassland supplies, besides the necessary amounts for cattle breeding and milk cows, biogas fermenters, with silage juice going to the Green Biorefinery although at reduced rates compared to the scenario 1A (as much silage goes to husbandry). Biogas is used for CHP, generating electricity for regional consumption (which can be covered if photovoltaic, wind power and hydro power are utilised to capacity). Part of the biogas is again cleaned and used for transport fuel however this part is considerably lower than in scenario 1A. District heating uses excess heat from these CHP-plants with the shortfall filled by using miscanthus grown on fields in heating plants. Food for regional consumption can be supplied by local agriculture.

The use of waste wood, fat from slaughtering is the same as in the optimum scenarios, buildings again are insulated as much as possible. Transport fuel however cannot be supplied in an amount to meet transport needs at the current level.

Supplying Linz scenario 3

Supplying the urban centre of Linz with food (cereals as well as meat) at a slightly higher level than today of course requires land that is then not available for either energy resources or other ways of utilisation that increase the added value in the region. Although regional heat demand can be met by using wood and employing insulation to increase energy efficiency of buildings, neither electricity demand nor transport energy can then be supplied in the amount to meet current levels of consumption. In general this scenario calls for a similar technological structure like scenario 1A, albeit with lower capacities for biogas fermenters and Green Biorefineries as fewer resources may be allocated to energy and industrial use.

COMPARING SCENARIOS

The scenarios differ regarding the supply of energy for the region. Heat demand can always be met; however transport and electricity demand vary in their degree of regional supply as shown in Figure 2.

Regarding the economic parameters the scenarios differ widely, especially with respect to the ratio between investment and revenue. Figure 3 shows this for all scenarios.

Figure 3 shows that scenario 1B shows a slightly lower revenue than 1A (roughly 3%) however needs 16% less investment. Autarky requires almost the same investment as the optimal technology network however achieves only 63% of the revenue. Supplying a major urban centre with food decreases revenue dramatically, to below 60% of the optimum level. This scenario however also requires the lowest investment with only 70% the amount necessary for the optimum scenario.

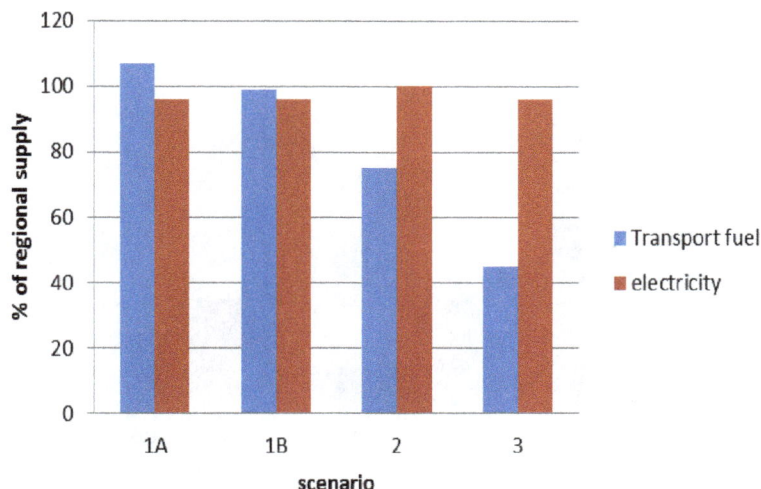

Figure 2. Regional coverage of transport and electricity demand in the scenarios

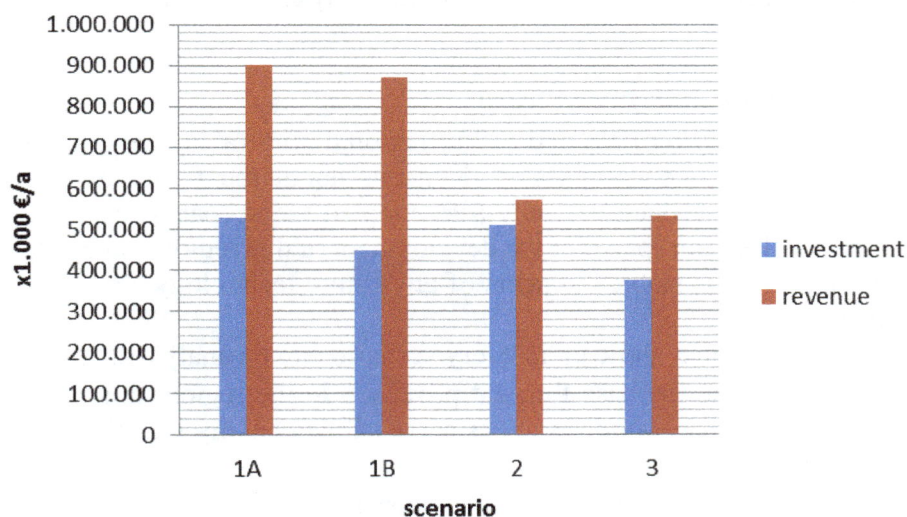

Figure 3. Investment (annualised with 10 years depreciation time) and yearly revenue for all scenarios

Concerning the ecological pressure, all scenarios presented here are reducing the ecological footprint against the status quo considerably as Figure 4 shows. This figure also shows the reduction in the ecological footprint for providing energy in the different scenarios using the different technological pathways defined by PNS optimisation.

Figure 4 shows that as all scenarios change the energy system of the region mostly towards renewable resources, overall ecological pressure of the region is reduced to a third (scenario 1 and 3) and even a quarter (scenario 2). All energy services provided by regional resources show massively reduced ecological footprints, with heat at only 20 % of the current status in all scenarios. The differences in the footprint of fuel are mostly due to the percentage of cleaned biogas used in the scenario, with scenario 1A showing a relatively high (but still much reduced) footprint for this energy form. By and large, autarky shows low ecological footprints. That has to be contrasted with the low economic performance of this scenario.

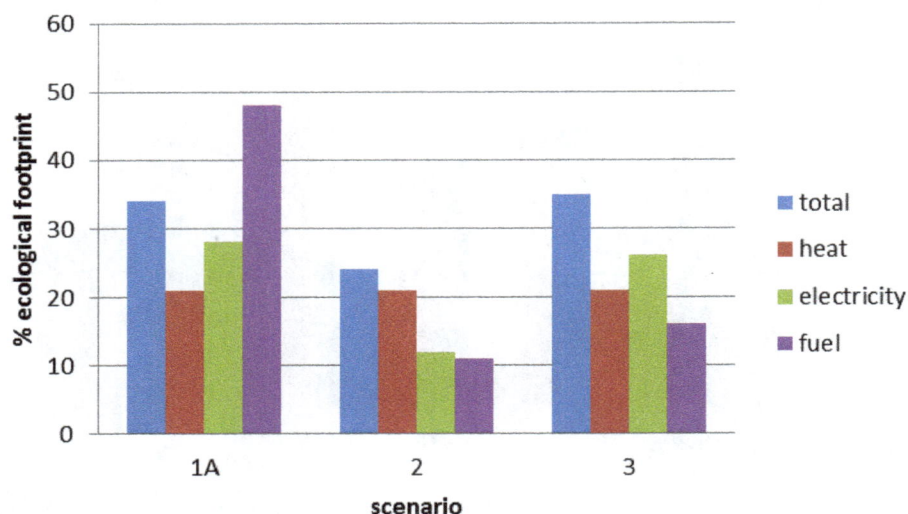

Figure 4: Comparison of ecological pressures using the SPI evaluation of all scenarios against the status quo (footprint for energy referring only to regional provision)

What regional decision makers can learn

As stated earlier, applying PNS and SPI to regional renewable energy systems must be seen in the context of encompassing decisions about the future of regions. The scenarios presented here (which are only a fraction of the scenarios calculated in this process) delimit the decision space for the development of this region and provide insight into the choices as well as stable elements of any future technology structure based on regional resources.

First stable elements that show up in every scenario can be analysed:

- Wood will become the base for heat provision in the region; this means that all measures to mobilise wood resources and establish energy logistics for wood are safe decisions for the region;
- District heating should be developed to capacity;
- PV as well as wind and hydro power should be developed to capacity;
- Insulating all buildings to low energy standards is necessary to gain energy efficiency;
- Biogas mobility shows great potential in all scenarios; this means that logistics for this form of fuel as well as measures to increase the car fleet that may use bio-methane as fuel are safe decisions for the future.

As interesting as the stable elements are the stark choices that the scenarios reveal. Amazingly enough, the future of the Mühlviertel critically depends on the utilisation of grassland and only to a minor degree on all other land resources. There is a choice to orient the region towards energy export and industrial utilisation of renewable resources (scenario 1A) or intensify marketing of existing agricultural products, in particular beef (scenario 1B). Both require major efforts to open new markets and to build up the necessary infrastructure and marketing structure. Whereas a focus on energy and industrial utilisation promises the highest revenues it also requires the highest investment.

Autarky as well as supply of a nearby urban centre will diminish revenue for the region considerably. Autarky in particular couples low revenues with high investment requirements.

All scenarios show much lower ecological pressures than the status quo, with the lowest overall environmental impact exerted by autarky. The environmental pressure for heat will be reduced to a fifth of the current level and stays relatively constant in all scenarios as the way heat is provided is stable throughout the scenarios. Both fuel and electricity footprints vary considerably, depending on the different pathways for their provision associated with the scenarios.

CONCLUSION

Regions will become major decision levels for the energy change necessary in the 21st century. As regional resources as well as demands are quite diverse, technological solutions will have to be adapted to the individual regional context. Utilising renewable resources to gain maximum regional revenue while exerting minimum ecological pressure will always require technological systems rather than single technologies, taking into account the framework and boundary conditions resulting from the ecological, logistical, economical and societal aspects of utilising renewable resources as discussed in this paper.

Implementation of radically new technological systems that entail major changes in business models and logistics require careful and participatory planning processes involving actors that have in many cases not co-operated before. This needs efficient tools that allow for systemic optimisation while providing insights into the long term choices to be taken. Adapting process synthesis and ecological process evaluation to the regional case can help to provide decision makers with comparable scenarios that will guide the planning process.

The case study shows that these tools will lead to a much clearer picture about the specific challenges for regional development when introducing renewable energy systems. It also shows clearly that using regional renewable resources lead to considerable chances for increasing regional revenue while cutting ecological impact dramatically.

ACKNOWLEDGEMENTS

The research presented here was carried out under the project "Ressourcenplan Mühlviertel" funded by the Austrian Climate and Energy Fund and carried out within the programme "NEUE ENERGIEN 2020" (grant Number 821845). We thank our fellow researchers Martin Luger, Robert Tichler and Horst Steinmüller from the Energieinstitut an der Johannes Kepler Universität Linz GmbH, Gerald Lutz and Reinhold Priewasser as well as Robert Robeischl from the Regionalmanagement Oberösterreich.

REFERENCES

1. IEA *Resources to Reserves-Oil and Gas Technologies for the Energy Markets of the Future*, (2005), www.iea.org [Accessed: 20-Jan-2012]
2. Schindler, J., Zittel, W., Fossile Energiereserven und mögliche Versorgungsengpässe. Studie für den Deutschen Bundestag, LB-Systemtechnik, Ottobrunn, Germany, 2000
3. Narodoslawsky, M., Niederl, A., Halasz, L., *Utilising renewable resources economically: new challenges and chances for process development*, Journal of

Cleaner   Production,   16/2,   pp   164-170,   2008.

4. Lafferty W. (Ed.) *Sustainable Communities in Europe*. Earthscan, London, 2001
5. Lafferty, W., Narodoslawsky, M. (Ed.), *Regional Sustainable Development in Europe: The Challenge of Multi-Level Co-operative Goverance*. ProSus, University of Oslo, 2002
6. O'Riordan, T. *Globalism, Localism & Identity*. Earthscan, London, 2001
7. Narodoslawsky, M., Stoeglehner, G., *Planning for Local and Regional Energy Strategies with the Ecological Footprint*. J Environ Pol Plan, 12(4), pp 363-379, 2010.
8. Stöglehner, G., Niemetz, N., Kettl, K.-H., *Spatial dimensions of sustainable energy systems: new visions for integrated spatial and energy planning*. Energy, Sustainability and Society, 1, 1-9; ISSN, pp 2192-0567, 2011.

9. Friedler, F., Varga, JB., Fan, LT., Decision-mapping: a tool for consistent and complete decisions in process synthesis. Chemical Engineering Science Vol. 50, pp 1755-1768, 1995.
10. Gwehenberger G., Narodoslawsky M., Liebmann B., Friedl A., Ecology of scale versus economy of scale for bioethanol production, Biofuels, Bioproducts and Biorefining, 1 (4), pp 264 – 269, 2007.
11. Halasz L., Gwehenberger G., Narodoslawsky M., *Process synthesis for the sugar sector - computer based insights in industrial development* in: V. Plesiu (Ed.): Computer Aided Chemical Engineering, Vol 24, pp 431-436, 2007
12. P-Graph homepage, http://www.p-graph.com, [Accessed: 20-Jan-2012]
13. SPI homepage, http://spionexcel.tugraz.at, [Accessed: 20-Jan-2012]
14. Narodoslawsky, M., Niederl, A., *Sustainable Process Index,* Renewable-Based Technology: Sustainability Assessment, Chapter 10, Ed: Dewulf, J.; van Langhove, H., John Wiley & Sons, 2005
15. Krotscheck C., Narodoslawsky M., *The Sustainable Process Index A new dimension in ecological evaluation*. Ecological Engineering 6, pp 241-258, 1996.

16. Fußabdrucksrechner für die Landwirtschaft http://www.fussabdrucksrechner.at, [Accessed: Jan-2012]
17. Nachhaltig   Wirtschaften   (2006), http://www.nachhaltigwirtschaften.at/results.html/id3028, [Accessed: 20-Feb-2012]
18. Steinmüller H., et al., Endbericht des Projektes Mühlviertler Ressourcenplan, Austrian Climate Fund, 2012, in press

12

Estimation of In-Situ Groundwater Conditions Based on Geochemical Equilibrium Simulations

Toshiyuki Hokari[*1], *Teruki Iwatsuki*[2], *Takanori Kunimaru*[2]
[1]Shimizu Corporation, Institute of Technology, Japan
e-mail: toshiyuki.hokari@shimz.co.jp

[2]Tono Geoscientific Research Unit
Japan Atomic Energy Agency, Japan

ABSTRACT

This paper presents a means of estimating in-situ groundwater pH and oxidation-redox potential (ORP), two very important parameters for species migration analysis in safety assessments for radioactive waste disposal or carbon dioxide sequestration. The method was applied to a pumping test in a deep borehole drilled in a tertiary formation in Japan for validation. The following application examples are presented: when applied to several other pumping tests at the same site, it could estimate distributions of the in-situ groundwater pH and ORP; applied to multiple points selected in the groundwater database of Japan, it could help estimate the in-situ redox reaction governing the groundwater conditions in some areas.

KEYWORDS

Groundwater, Water conditions, In-situ, pH; Oxidation-redox potential (ORP), Geochemical calculation, Mineral-water interaction

INTRODUCTION

Geochemical characteristics of deep groundwater are essential information for safety assessments for the geological disposal of radioactive wastes [1], and the sequestration of carbon dioxide [2], one of the known greenhouse gases, because groundwater chemistry could affect migration of the species included in disposal wastes. In order to facilitate the smooth advance of the above disposal projects, it is necessary to investigate the geochemical characteristics economically across a wide area extending over several kilometres.

Existing investigations of groundwater chemistry so far have involved drilling a borehole, purging the drilling mud, pumping up the groundwater, sampling it at the surface and conducting analyses in the laboratory. At potential disposal locations (hereinafter referred to as in-situ), the groundwater is generally under high pressure to dissolve gases, i.e. carbon dioxide, and is in a reduced condition. When pumped up to the surface, it could be degassed with depressurisation to increase its pH and it could be oxidised by contact with the atmosphere to increase its oxidation redox potential (ORP) [3-6].

In order to procure quality data on the pH and ORP of the deep groundwater, it is recommended to measure them in-situ [7, 8], and some apparatus has been developed for

---

in-situ groundwater measurement and sampling. One problem is that since in-situ measurement takes longer and is more expensive, it is difficult to set up a network of measurements consisting of many test intervals in boreholes. A realistic solution is considered as follows: (1) perform not only in-situ measurements, but also the existing ones; (2) develop a method for estimating the in-situ pH and ORP using existing data in comparison with in-situ data; (3) estimate the in-situ values of test intervals where in-situ measurements are not conducted; (4) economically obtain data on the in-situ pH and ORP across a wide area.

This paper suggests a method for estimating the in-situ pH and ORP on the basis of existing data and chemical equilibrium analysis. Moreover, as application examples, the following is estimated: distributions of the in-situ pH and ORP at a site in Japan; a predominant redox reaction governing the in-situ groundwater conditions in use of the groundwater database of Japan.

ESTIMATION METHOD

In order to understand the evolution mechanism of groundwater chemistry, it is convenient to calculate speciation of elements in some environments with a thermodynamic code for geochemical modelling. The geochemical code enables calculations of species activities, concentrations and saturation indices in water on the basis of the mass balance law and the mass action law with a thermodynamic database that includes mass action constants. This paper employs one of the open codes, called PHREEQC [9], developed by the U.S. Geological Survey to estimate in-situ pH and ORP. There are several codes for geochemical modelling other than PHREEQC, which are detailed in the following websites [10].

Since the details of PHREEQC were presented by Parkhurst and Appelo (1999) [9], only a summary of PHREEQC is given here. It is designed to perform a wide variety of low-temperature aqueous geochemical calculations on the basis of an ion-association aqueous model. In order to estimate the in-situ water conditions on the basis of the existing surface data, of the many geochemical calculation capabilities of PHREEQC, speciation and batch reactions with gas at equilibrium are focused on here. Liquid phase interactions with the surrounding solid phase are considered later in the section of RESULT AND DISCUSSION. It uses the mole balance Equation (1), the mass action Equations (2), (3), and the activity coefficient expression including the Davies Equation (4) or the extended Debye-Huckel Equation (5) [11] to calculate the activities, concentrations and saturation indices of the species in solution.

The mole balance equation of an element m is expressed:

$$\sum_{i}^{N_{aq}} b_{m,i} n_i + \sum_{g}^{N_g} b_{m,g} n_g = const. \tag{1}$$

where N_{aq} is the number of aqueous species, and N_g is the number of gas-phase species. The moles of each species in the system are represented by n_i for aqueous species and n_g for gaseous species. The moles of element m per mole of each species are represented by $b_{m,\,i}$ for aqueous species and $b_{m,\,g}$ for gaseous species.

The mass action equations can lead to the total moles of an aqueous species i and a gaseous species g:

$$n_i = \frac{K_i W_{aq}}{\gamma_i} \prod_{m}^{M_{aq}} a_m^{c_{m,i}} \tag{2}$$

$$n_g = \frac{N_{gas}}{P_{total}K_g} \prod_m^{M_{aq}} a_m^{c_{m,g}} \tag{3}$$

where n is the moles, K is the mass action constant, a_m is the activity of master species m, M_{aq} is the total number of aqueous master species, c_m is the stoichiometric coefficient of master species m, W_{aq} is the mass of solvent water in an aqueous solution, γ is the activity coefficient, N_{gas} is the total moles of gas, P_{total} is the total pressure, and subscript i, g represents a solutions species and a gaseous species, respectively.

Activity coefficient γ of aqueous species i is defined with the Davies Equation (4) or the extended Debye-Huckel Equation (5):

$$\log \gamma_i = -Az_i^2 \left(\frac{\sqrt{\mu}}{1+\sqrt{\mu}} - 0.3\mu \right) \tag{4}$$

$$\log \gamma_i = \frac{Az_i^2 \sqrt{\mu}}{1 + Ba_i^0 \sqrt{\mu}} + b_i \mu \tag{5}$$

where z_i is the ionic charge of aqueous species i, μ is the ionic strength of solution, A and B are constants dependent only on temperature, and a_i^0 and b_i are ion-specific parameters fitted from mean-salt activity-coefficient data.

The initial input to PHREEQC was the following analysis data on the groundwater pumped up to the surface: the temperature, pressure (1 atm), pH, ORP, main species concentrations, if there were free gases found, the gas/water ratio, and content of each gas. The groundwater conditions under the in-situ pressure and temperature were computed with PHREEQC on the basis of the initial solution. With increasing pressure, the free gases in the solution were expected to be all solved. The equations below the bubble point were expected to differ from those above it according to the presence or dissolution of the gases. In order to estimate the bubble point, a stepwise computation was applied from the surface pressure and temperature conditions to the in-situ one. If the in-situ mineral information was available, the effects of the mineral on pH or ORP were to be considered and added to the simulation result with PHREEQC.

GEOCHEMICAL PUMPING TEST

The estimation method for the in-situ water conditions was applied to a geochemical pumping test for validation. It was performed by the Japan Atomic Energy Agency (JAEA) in the deepest borehole drilled in the course of the Horonobe Underground Research Laboratory Project (Horonobe URL project) [12]. The details of the geochemical pumping test are presented by Hokari and Kunimaru [13], and a summary of the test is given here.

Figure 1 represents the location of the Horonobe URL project, which has been implemented by JAEA in Hokkaido, Japan, with the locations of the investigation boreholes and surrounding geology. The test was conducted at approximately 600 m deep intervals in a vertical borehole named HDB-11. Analysis of the geological column of HDB-11 revealed that it consists of a diatomaceous mudstone Koetoi formation from the surface to a depth of 460 m, and a hard shale Wakkanai formation below 460 m.

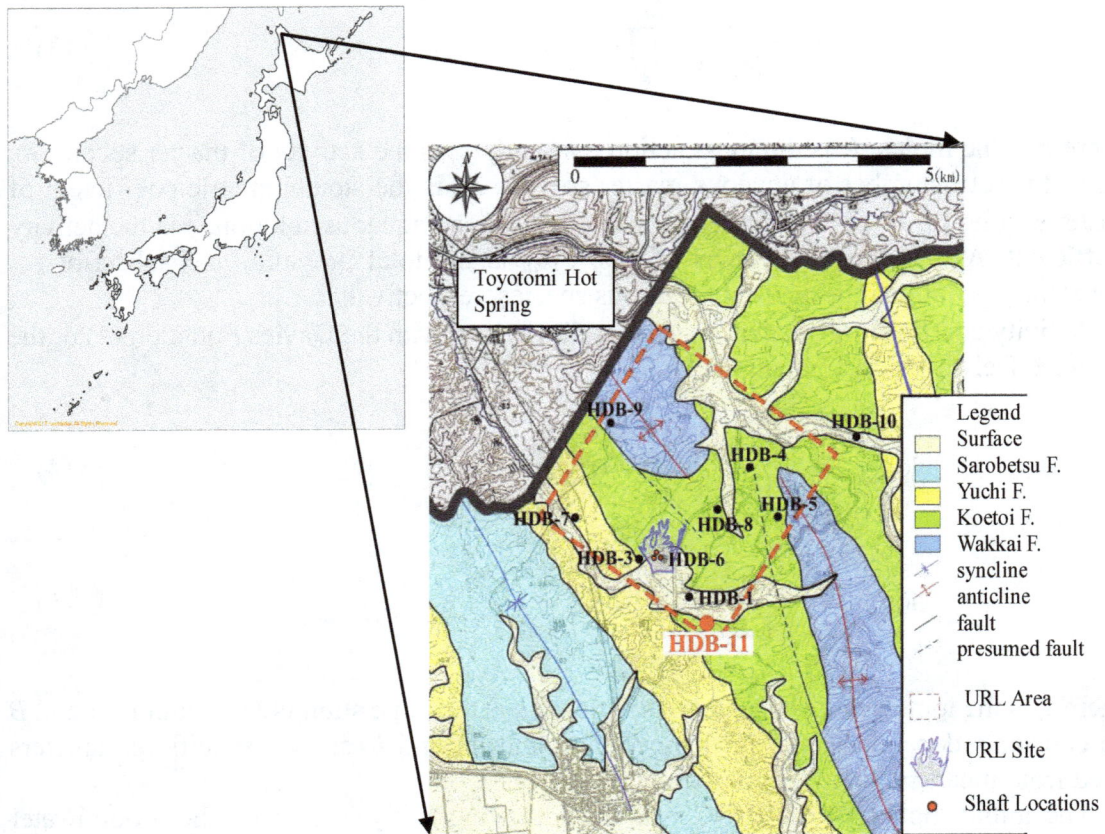

Figure 1. Location and Geology of Horonobe URL Project

Groundwater at Horonobe site was deduced from the results of some investigations to have evolved as follows: Whereas in the vicinity of the surface, a fresh groundwater of meteoric water origin with a fewer solute contents prevailed, there was a saline groundwater of sea water origin with more solutes in the depths; The present groundwater around the site has been formed by mixing of the above two end waters [12].

Prior to the geochemical pumping test, a hydraulic pumping test was conducted at the same interval of 606 ~ 644 m depth. The following hydraulic properties were obtained:

hydraulic head	hydraulic conductivity	specific storage
GL + 5.3m	2.3×10^{-8} m/s	4.3×10^{-5}/m

The geochemical test consisted of pumping the groundwater, monitoring of physical-chemical parameters including water pressure (only in-situ), pH, temperature, dissolved oxygen (DO), electrical conductivity (EC) and ORP, and in-situ water sampling which had the capability of maintaining the in-situ water pressure with stainless steel containers. The parameters were monitored both in-situ and on the surface except for the pressure. Figure 2 presents an outline of the geochemical pumping test apparatus.

In-situ measurements were made using an OCEAN SEVEN 303 constructed by IDRONAUT, and the surface measurements used WM-50EG for pH, EC and temperature, IM-55G for ORP (Pt electrode) and DO-55G for DO, all made by Toa DKK and installed in the flow-through cell.

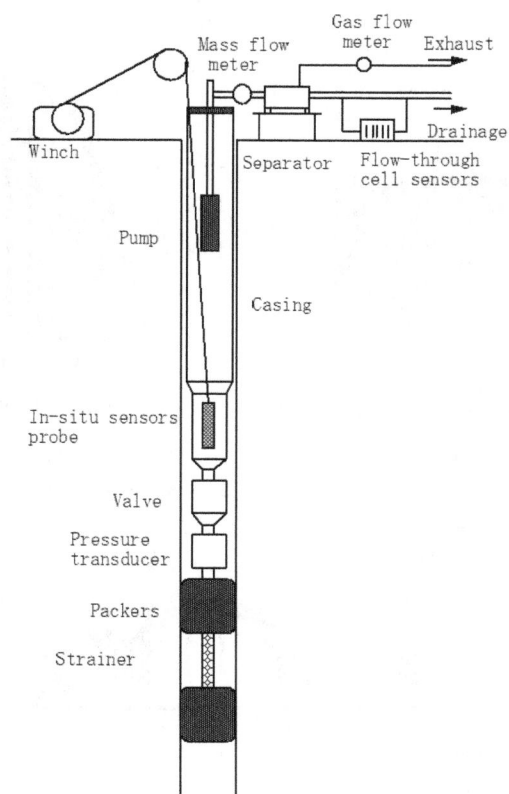

Figure 2. Schematic Drawing of the Geochemical Pumping Test Apparatus

The geochemical test procedure is summarised as follows: 1) installation of the pumping apparatus at the test interval in the borehole, 2) packer inflation, 3) pore pressure measurement, 4) installation of in-situ sensor probe, 5) installation of the pump and wellhead assembly, 6) installation of the flow-through cell sensors on the surface, 7) pumping and groundwater monitoring, 8) in-situ groundwater sampling after removal of the in-situ sensors and the pump, 8) six repetitions of the monitoring and water sampling, 9) packer deflation.

Some measurements of the physical-chemical parameters in-situ and on the surface are shown in Figure 3. The horizontal axis represents the ratio of pumped-up groundwater volume to the test interval one of 0.73 m^3, and the vertical axis shows pH, EC and ORP_SHE, respectively. ORP_SHE indicates an ORP value relative to the standard hydrogen electrode which was converted from the ORP measurement. EC measurements were corrected and adjusted to 25 $^{\circ}$C [14].

IN-SITU and GROUND in the legends for Figure 3 show measurements monitored in-situ and on the surface, respectively. The pumping procedure included six sets of pumping and monitoring periods and a bottle sampling, the final set of which was the longest. Since the in-situ sensor probe was retrieved from the borehole during the bottle sampling, the measurement curves were intermittent in Figure 3. It is observed that the in-situ measurements differed from the surface ones in pH and ORP, whereas the in-situ EC measurement was in good agreement with that at the surface.

The pore pressure was measured as approximately 6.0 MPa at an initial equilibrium and was stably maintained at 5.8 MPa during pumping. The in-situ temperature was stable and measured approximately 35 $^{\circ}$C during the period of the test over around 20 days. DO was not detected in-situ, but a trace amount was detected at the surface.

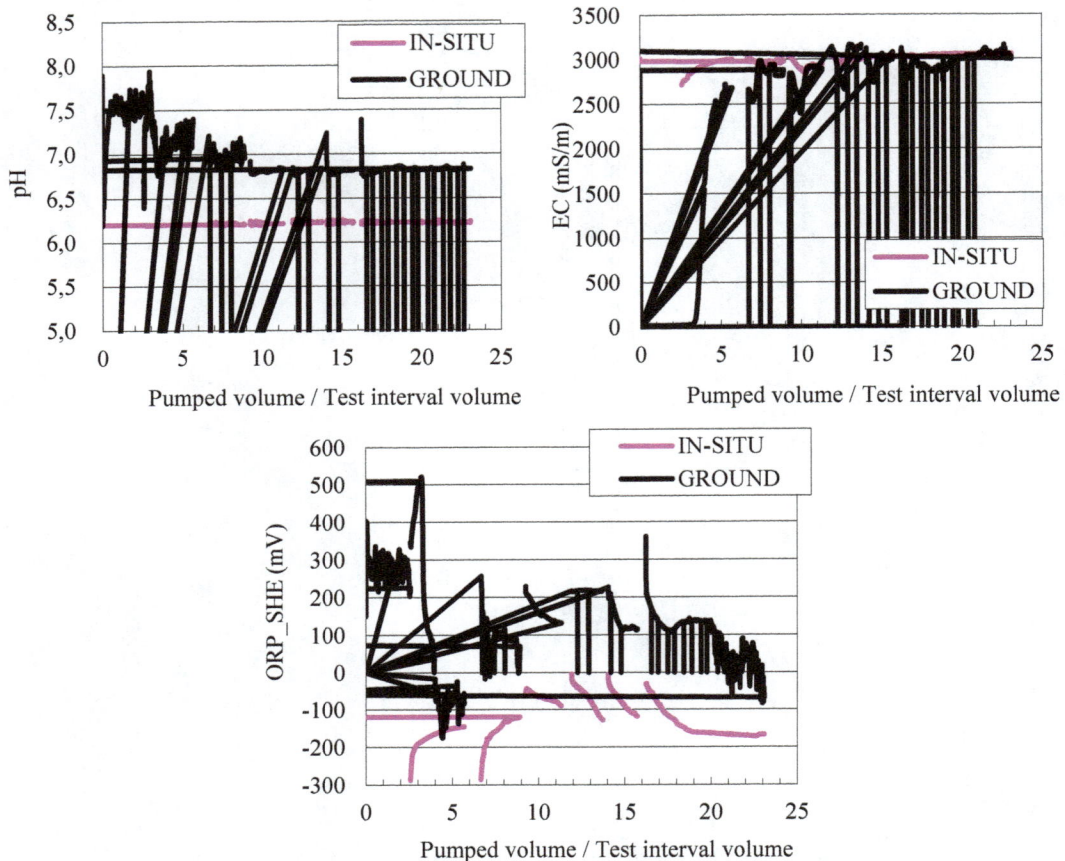

Figure 3. Monitoring Result of pH, EC and ORP

The flow-through cell sensors were set between a separator and a drain, and measured the groundwater before any contact with the atmosphere. Iwatsuki et al. [6] argued that since the flow cell includes free gases released from the groundwater, which is supposed to be higher than the atmosphere in pressure, it would be less possible to oxidise the groundwater owing to the air intrusion. Grenthe et al. [3] and Gascoyne [5], however, conclude that it is difficult to completely prevent air intrusion into the flow-cell system at the surface. As this pumping test detected DO at the surface, air intrusion could occur in the groundwater pumped up, which could explain the ORP difference between the in-situ and the surface results.

The in-situ groundwater was sampled using high-pressure stainless steel bottles that could maintain the in-situ pressure. Many aqueous species were analysed at the laboratory after the high- pressure bottle samples were depressurised at the site in the air. To analyse a redox couple of Fe^{2+}/Fe^{3+}, the high-pressure water sample was depressurised in an inert atmosphere in the laboratory. In order to measure the gas/water ratios, high-pressure samples were released into a high-pressure vessel of a given volume which had been evacuated so that measurement of the gas pressure may result in a volume of gas being released. The gas contents were analysed using gas chromatography after the gas sampling in the pressure vessel. The aqueous species concentrations and the free gas contents are compiled in Table 1 and Table 2. The gas/water ratios were measured as approximately 1.5. Flame photometry was used for analysis of Na^+ and K^+. Inductively coupled plasma optical emission spectroscopy was used for Ca^{2+}, Mg^{2+}, Mn^{2+} and dissolved Si. Absorptiometry was used for NH_4^+, Fe^{2+}, Fe^{3+} and S^{2-}. Ion chromatography was used for Cl^-, SO_4^{2-}, F^-, Br^-, I^- and NO_3^-. Titration was used for HCO_3^- and CO_3^{2-}.

Table 1. Contents of the groundwater aqueous species

Cation Concentration [mg/L]								
Na^+	K^+	Ca^{2+}	Mg^{2+}	Mn^{2+}	Fe^{2+}	Fe^{3+}	NH_4^+	Si
6600	140	250	170	0.01	2.3	<0.05	200	27
Anion Concentration [mg/L]								
Cl^-	HCO_3^-	CO_3^{2-}	SO_4^{2-}	S^{2-}	F^-	Br^-	I^-	NO_3^-
10000	2200	0	<0.2	<0.1	0.1	78	29	0.1

Table 2. Free gas contents

O_2 [%]	N_2 [%]	CO_2 [%]	Ar [%]	CH_4 [%]	C_2H_6 [%]	H_2S [ppm]
0.0	0.0	26.3	0.04	73.6	0.015	0.0

The stable in-situ pressure during the pumping test implied that no gases would be released from the in-situ groundwater. The in-situ groundwater would release one and a half times the volume of gas as the water on the surface, which consists of 75% CH_4 and 25% CO_2. This could lead to differences in the pH and ORP between in-situ and surface samples.

RESULT AND DISCUSSION

According to Figure 3, the pH measurements were stable both in-situ and on the surface. Whereas the in-situ ORP measurement was observed to be stable near the end of the pumping test, the surface measurement was found not to reach a steady state, but to be around -50 ~ -150 mV. The pH and ORP measurements and the PHREEQC computation estimates were as follows:

pH measurements on the surface and in-situ	6.80	6.20
pH estimates on the surface and in-situ	6.80	6.29
ORP measurements on the surface and in-situ (mV)	-50 ~ -150	-166
ORP estimates on the surface and in-situ (mV)	-198	-210

As for pH, the surface estimate was in good agreement with the measurement and there was a difference of 0.1 between the in-situ measurement and the estimate. As for ORP, since the surface measurement did not attain a steady state, it was not useful to compare the surface measurement with the estimate. The ORP in-situ estimate indicated a more reduced state than the measurement by 45 mV.

The HDB-11 borehole investigation conducted a rock core analysis as well as the pumping tests, so in-situ mineral information may be available [15]. Mineral effects on groundwater conditions are considered here and added to the PHREEQC computation results.

The groundwater pH would vary according to a change in the carbonate acid equilibrium caused by CO_2 degassing with depressurisation. The deep groundwater is expected to be in equilibrium with rock forming minerals in the surrounding formations due to very prolonged residence. The pH would also be governed by carbonate mineral equilibrium such as calcite in-situ, which was identified by the core analysis. A reaction of carbonic acid and water and a pH equation are given in Equation (6). A reaction of calcite dissolution into the water and a pH equation are given in Equation (7). Here, K represents a mass action constant, the activity of an ion is denoted by square brackets, i.e. $[Ca^{2+}]$, and subscript (aq) presents the aqueous species. Two values of log K in each equation correspond to those on the surface and under in-situ conditions.

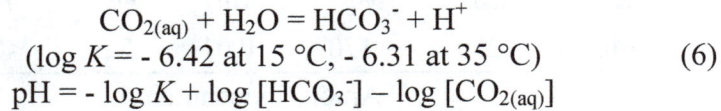

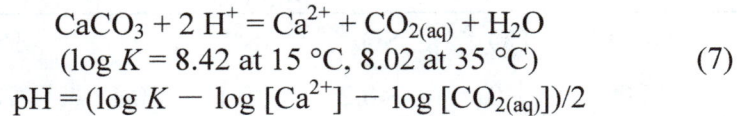

$$CO_{2(aq)} + H_2O = HCO_3^- + H^+$$
$$(\log K = -6.42 \text{ at } 15 \text{ °C}, -6.31 \text{ at } 35 \text{ °C}) \tag{6}$$
$$pH = -\log K + \log [HCO_3^-] - \log [CO_{2(aq)}]$$

$$CaCO_3 + 2 H^+ = Ca^{2+} + CO_{2(aq)} + H_2O$$
$$(\log K = 8.42 \text{ at } 15 \text{ °C}, 8.02 \text{ at } 35 \text{ °C}) \tag{7}$$
$$pH = (\log K - \log [Ca^{2+}] - \log [CO_{2(aq)}])/2$$

Since PHREEQC computed the above species activities in Equations (6) and (7), the relationship curves between $CO_{2(aq)}$ and pH are given with the measurements and the estimates of pH in Figure 4. A carbonate equilibrium curve is calculated on the surface condition of 15 °C, and a calcite curve on the in-situ condition of 35 °C. The surface pH values are plotted on the carbonate acid equilibrium curve, which is close to the in-situ estimate, and the in-situ measurement is plotted on the calcite equilibrium. The in-situ estimate is actually plotted on the carbonate acid equilibrium curve of 35 °C. It is interpreted that whereas the pH of the groundwater pumped up to the surface will be governed by the carbonate acid equilibrium, the in-situ pH will be governed by the calcite equilibrium. PHREEQC is found to simulate the pH values of the groundwater on the basis of the carbonate acid equilibrium. It is understood that a consideration of the in-situ mineral reaction added to the computation result by PHREEQC would produce a more accurate estimate of the in-situ pH.

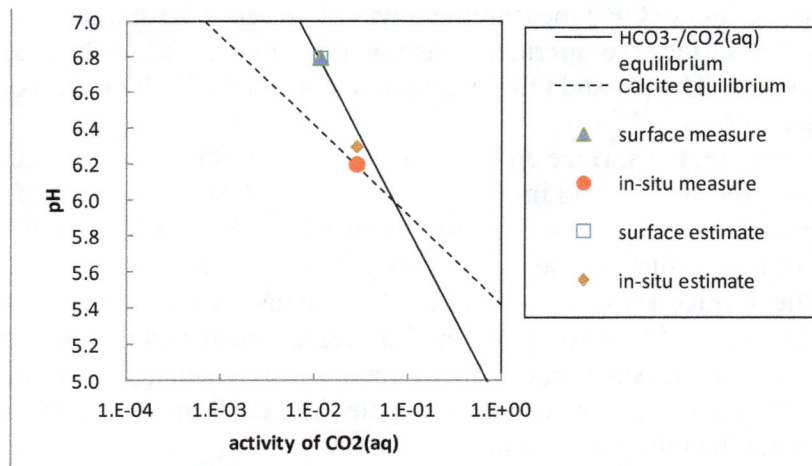

Figure 4. Analytical result of the pH measures and estimates

While assuming redox reactions including the aqueous species analysed and the minerals identified [16, 17], the ORP of each reaction in Figure 5 is calculated thermodynamically by the Nernst equation with standard potential, $E0$, on the basis of the activities of the PHREEQC computation results. Figure 5 shows the calculated ORP-pH relationships with the ORP measurements and the PHREEQC estimates. Although sulphate and sulphide were not detected in the groundwater, pyrite and gypsum were identified from the core analysis. Sulphate and sulphide species are expected to be contained in the water at levels below the detection limit for this analysis of 0.1 mg/L, respectively. The ORP values in Figure 5 are calculated on the assumption that the groundwater should include total sulphur of 1×10^{-7} M, which is approximately a tenth of the detection limit. Some curves running on and near the measurements represent that the in-situ redox condition is governed by the sulphide/sulphate redox reaction or the goethite/siderite reaction, and is also strongly affected by the pyrite/sulphate reaction. This suggests that in-situ, the sulphate/sulphide reaction occurs predominantly after or at the same time as the sulphide supply due to dissolution of the pyrite and/or the iron-containing minerals of pyrite, siderite and ferric oxyhydroxide such as goethite are in equilibrium with each other. ORP behaviours vary widely according to the reactions involved as seen in Figure 5. Here, they could be divided into three patterns, one of which shows higher ORP values than the estimates and the measurements, one of which is lower than those values, and the other of which falls between them. The in-situ ORP is re-calculated on the basis of the latter type of reactions, because they are closest to the in-situ and surface measurements. The result is as follows:

ORP measurement	PHREEQC	sulphide/sulphate	goethite/siderite	pyrite/sulphate
-166 mV	-210 mV	-173 mV	-173 mV	-189mV

It is understood that a consideration of the in-situ mineral reactions could produce more accurate ORP estimates on the basis of the computation results by PHREEQC.

The estimation results for pH and ORP are compiled with the measurements obtained by the geochemical pumping test in Table 3. The geochemical equilibrium calculations could estimate the in-situ pH and ORP within a margin of error of approximately 0.1 and 50 mV on the basis of the analytical data on the groundwater and the free gases obtained from the pumping test. Furthermore, it is confirmed that as carbonate minerals, such as calcite, affecting the pH were identified in the rock core analysis, a consideration of the in-situ calcite reaction could allow a more accurate estimation of the in-situ pH through use of the activities of aqueous CO_2 and Ca^{2+} computed by PHREEQC. It is also confirmed that consideration of the in-situ redox reactions of the minerals such as pyrite and siderite, which were identified in the rock core analysis, could allow for more accurate estimation of the in-situ ORP in use of the activities of the species concerning the reactions computed by PHREEQC.

This estimation method is next applied to the several existing pumping tests that were conducted at the Horonobe site [18]. Figure 6 illustrates the depth distributions of the in-situ pH and ORP estimates at the site [19]. The in-situ pH estimates are more acidic than the surface measurements and have a tendency towards acidity to approximately 6.2 with depth. It is coincident with the pumping test results that more CO_2 gas was released from the deeper groundwater pumped up to the surface. The in-situ ORP estimates represent a more reduced condition than the surface measurements, and show stable reduction of approximately -200 mV at depths of below some 200 m. This is in agreement with a general tendency that the underground environment is reductive and stable in Japan [17].

Another application of the method is to groundwater data [21], and the database in Japan [22] for analysis of the in-situ conditions governing redox reactions. Approximately 70 points are selected to satisfy the PHREEQC input data conditions. Figure 7 illustrates the locations of the wells including the Dohoku area (black solid circle), Eastern Kanto district with Seki et al. data [21] (blue solid circle) and other areas (red solid circle). The Dohoku area includes the JAEA boreholes, petroleum wells and hot spring wells. Seki et al. data all comes from hot spring wells. The other area includes natural gas wells, coal mine boreholes, observation wells and hot spring wells. According to the intervals of the wells at which the groundwater was pumped up, the geology is quite different as follows: mudstone, shale, siltstone, sandstone, green tuff, tuffaceous breccia, welded tuff, granite, granodiorite, rhyolite and andesite. Redox reactions involve elements contained in the rocks and the groundwater, which could vary in valence states depending on the redox condition. At first, the elements of Fe, Mn, S and C are selected as redox relevant ones in the rocks, because they exist in greater content in the various rocks [23]. The relative contents of the elements are summarised in Table 4.

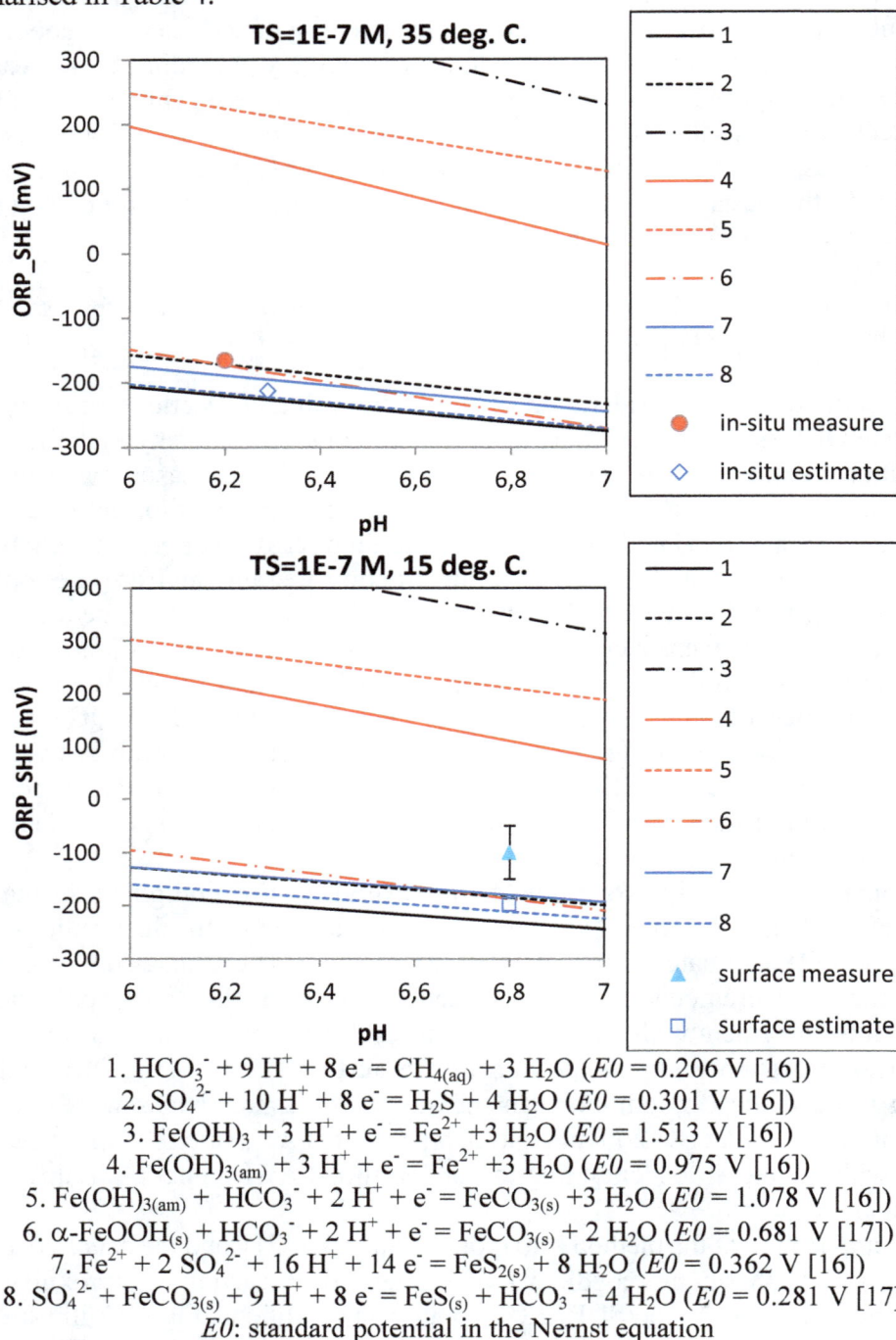

1. $HCO_3^- + 9\ H^+ + 8\ e^- = CH_{4(aq)} + 3\ H_2O$ ($E0 = 0.206$ V [16])
2. $SO_4^{2-} + 10\ H^+ + 8\ e^- = H_2S + 4\ H_2O$ ($E0 = 0.301$ V [16])
3. $Fe(OH)_3 + 3\ H^+ + e^- = Fe^{2+} + 3\ H_2O$ ($E0 = 1.513$ V [16])
4. $Fe(OH)_{3(am)} + 3\ H^+ + e^- = Fe^{2+} + 3\ H_2O$ ($E0 = 0.975$ V [16])
5. $Fe(OH)_{3(am)} + HCO_3^- + 2\ H^+ + e^- = FeCO_{3(s)} + 3\ H_2O$ ($E0 = 1.078$ V [16])
6. $\alpha\text{-}FeOOH_{(s)} + HCO_3^- + 2\ H^+ + e^- = FeCO_{3(s)} + 2\ H_2O$ ($E0 = 0.681$ V [17])
7. $Fe^{2+} + 2\ SO_4^{2-} + 16\ H^+ + 14\ e^- = FeS_{2(s)} + 8\ H_2O$ ($E0 = 0.362$ V [16])
8. $SO_4^{2-} + FeCO_{3(s)} + 9\ H^+ + 8\ e^- = FeS_{(s)} + HCO_3^- + 4\ H_2O$ ($E0 = 0.281$ V [17])
$E0$: standard potential in the Nernst equation

Figure 5. Analytical result of the ORP measurements and estimates

Table 3. Estimation results

Parameter	Location	Measurement	1st Estimate[1]	2nd Estimate[2]
pH	Surface	6.80	6.80	
	In-situ	6.20	6.29	6.20
ORP	Surface	-50~-150*	-198	
	In-situ	-166	-210	-173~-189

1) Estimates computed by PHREEQC
2) Estimates calculated on PHREEQC results and mineral reaction
* Provisional value as the measurement reached no steady state

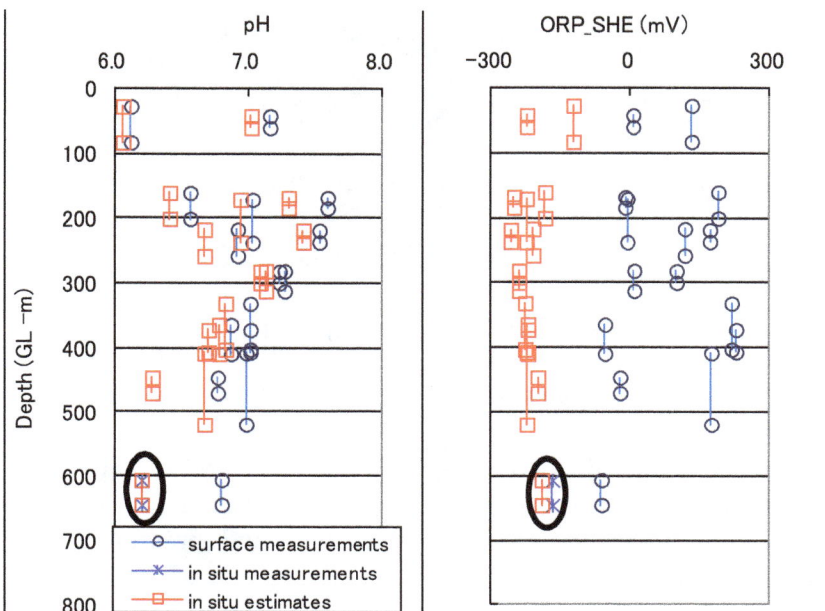

Figure 6. An estimation example of the in-situ pH and ORP distributions

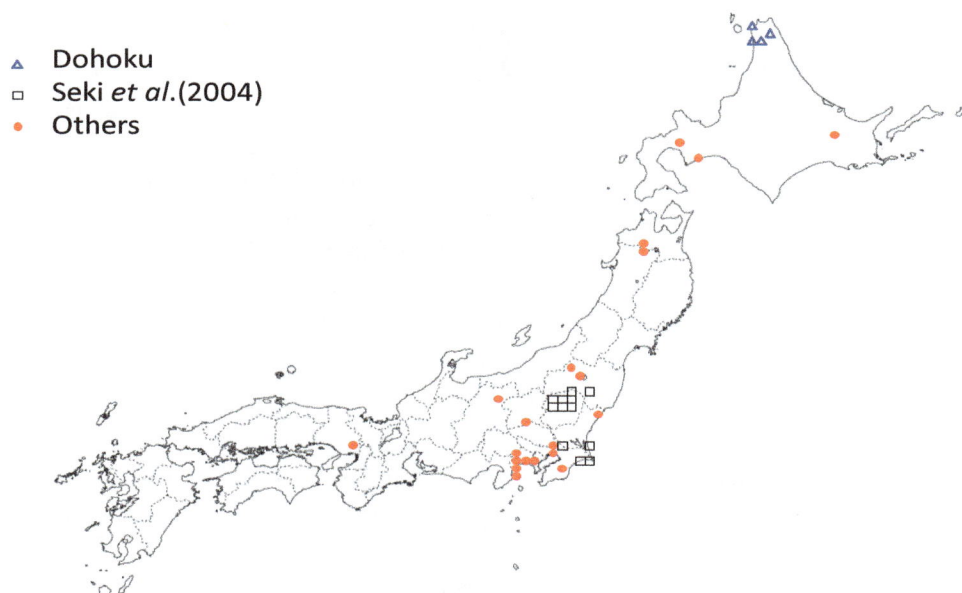

Figure 7. Locations of the wells for the in-situ redox reaction estimation

Table 4. Main elements concerning the redox state and relative contents (modified from [23])

Igneous rocks		Fe > Mn > S > C
	Sandstone	Fe > C > S > Mn
Sedimentary rocks	Shale	Fe > C > S > Mn
	Carbonates	C > Fe > S > Mn

The selected groundwater data seldom includes Mn, and the following redox reactions are assumed with the elements of Fe, S and C in the in-situ groundwater and the rocks.

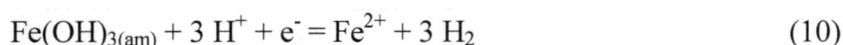

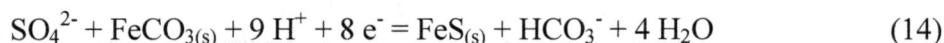

$$SO_4^{2-} + 10\,H^+ + 8e^- = H_2S_{(aq)} + 4\,H_2O \tag{8}$$

$$Fe(OH)_3 + 3\,H^+ + e^- = Fe^{2+} + 3\,H_2O \tag{9}$$

$$Fe(OH)_{3(am)} + 3\,H^+ + e^- = Fe^{2+} + 3\,H_2 \tag{10}$$

$$Fe(OH)_{3(am)} + HCO_3^- + 2\,H^+ + e^- = FeCO_{3(s)} + 3\,H_2O \tag{11}$$

$$\alpha\text{-}FeOOH_{(s)} + HCO_3^- + 2\,H^+ + e^- = FeCO_{3(s)} + 2\,H_2O \tag{12}$$

$$Fe^{2+} + 2\,SO_4^{2-} + 16\,H^+ + 14\,e^- = FeS_{2(s)} + 8\,H_2O \tag{13}$$

$$SO_4^{2-} + FeCO_{3(s)} + 9\,H^+ + 8\,e^- = FeS_{(s)} + HCO_3^- + 4\,H_2O \tag{14}$$

The predominant reaction of all the above could be revealed with thermodynamic analysis using the Gibbs reaction energy. The energy was calculated for each reaction for each groundwater data under the in-situ temperature and pressure conditions. The probability of each reaction is as follows:

Reaction	(8)	(9)	(10)	(11)	(12)	(13)	(14)
Probability [%]	0	0	34	1	12	36	17

The above probability means, for example, that Reaction (13) is the most likely to occur in 36% of all the data. The sulphate/ferrous sulphide mineral reactions are estimated to be predominant in more than half of the data, the ferrous ion/ferric oxihydroxide reaction predominates in 34% of data, and the siderite/ferric oxihydroxide reaction prevails in 12%. In other words, the redox reactions of the ferrous sulphide minerals are estimated to govern the in-situ groundwater conditions. The in-situ pH and ORP estimates are analysed on the basis of Reaction (13), as shown in Figure 8. Most of the estimates are plotted on an equilibrium curve between pyrite and sulphate. It is deduced from the result that the redox state of the in-situ groundwater could be governed by the pyrite-sulphate reaction in some areas of Japan.

CONCLUSION

This study developed a means of estimating the in-situ pH and ORP, which are very important parameters affecting migration properties in the safety assessment of underground disposal facilities. This was applied to a geochemical pumping test for validation. The following application examples were also shown: When applied to several pumping tests in a given area, it could estimate distributions of the in-situ groundwater pH and ORP in the area; applied to a range of data on deep groundwater in a database of Japan, it could help estimate the in-situ redox reactions governing the groundwater conditions.

Since in-situ pH and ORP measurement is very expensive and time-consuming in the case of borehole investigations, this method could be utilised as follows: (1) The cost and time for the groundwater investigation is expected to be reduced by means of in-situ pH and ORP estimation by using data obtained from existing pumping tests; (2) A safety assessment prior to the site investigation is expected to be performed by means of the in-situ pH and ORP estimation on the basis of the existing groundwater database.

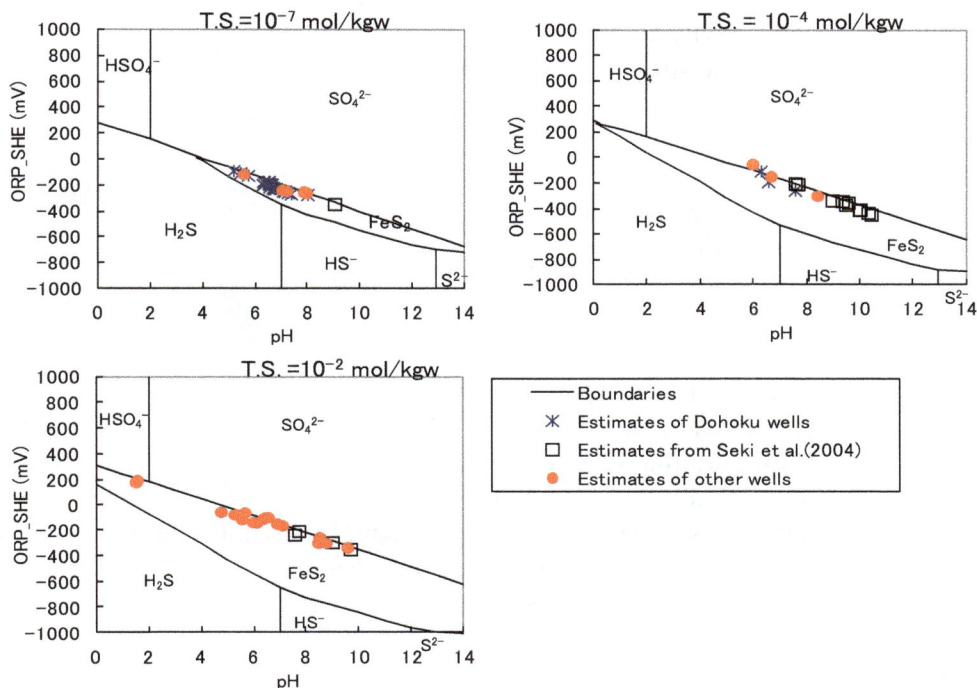

Figure 8. Estimation example of the in-situ redox reaction

NOMENCLATURE

A	-	the constant dependent only on temperature for the Davies Equation and the extended Debye-Huckel Equation
a_i^0	[m]	the ion-specific parameter fitted from mean-salt activity-coefficient data for the extended Debye-Huckel Equation
a_m	-	the activity of master species m
[a]	-	the activity of species a
B	[/m]	the constant dependent only on temperature for the extended Debye-Huckel Equation
b_i	-	the ion-specific parameter fitted from mean-salt activity-coefficient data for the extended Debye-Huckel Equation
b_m	-	the moles of element m per mole of species
c_m	-	the stoichiometric coefficient of master species m
$E0$	[V]	the standard potential in the Nernst equation
K	-	the mass action constant
M_{aq}	-	the total number of aqueous master species m
m	-	element
	-	master species

N_{aq}	-	the number of aqueous species in the system
N_g	-	the number of gas-phase species in the system
N_{gas}	[mol]	the total moles of gas
N	[mol]	the moles of species in the system
P_{total}	[atm]	the total pressure
W_{aq}	[kg]	the mass of solvent water in an aqueous solution
z	-	the ionic charge

Greek Letters

γ	-	the activity coefficient
μ	-	the ionic strength of solution

Subscripts

i	a solution species
g	a gaseous species

Abbrevations

DO	dissolved oxygen
EC	electrical conductivity
JAEA	Japan Atomic Energy Agency
ORP	oxidation-redox potential
ORP_SHE	an ORP value relative to the standard hydrogen electrode which was converted from the ORP measurement
URL	the Underground Rock Laboratory

REFERENCES

1. Nuclear Waste Management Organization of Japan (NUMO), Development of Repository Concepts for Volunteer Siting Environment, NUMO-TR-04-03, *NUMO*, 2004.

2. Xue, Z., Matsuoka, T., Lessons from the First Japanese Pilot Projects on Saline Aquifer CO2 Storage, *J. Geography*, Vol. 117, No. 4, pp 734-752, 2008.

3. Grenthe, I., Stumm, W., Laaksharju, M., Nilsson, A.C., Wikberg, P., Redox potentials and redox reactions in deep groundwater systems, *Chem. Geol.*, Vol. 98, No. 1-2, pp 131-150, 1992.

4. Gascoyne, M., Evolution of Redox Conditions and Groundwater Composition in Recharge-Discharge Environments on the Canadian Shield, *Hydrogeol. J.,* Vol. 5, No. 3, pp 4-18, 1997.

5. Gascoyne, M., Hydrogeochemistry, groundwater ages and sources of salts in a granitic batholiths on the Canadian Shield, south-eastern Manitoba, *Applied Geochem.*, Vol. 19, pp 519-560, 2004.

6. Iwatsuki, T., Morikawa, K., Hosoya, S., Yoshikawa, H., A notice for measuring physicochemical parameters (pH, ORP) of deep groundwater, *J. Jap. Assoc. Groundwater Hydrol.,* Vol. 51, No. 3, pp 205-214, 2009.

7. Ii, H., Horie, Y. Ishii, T., Shimada, J., Development of an apparatus to measure groundwater qualities in situ and to sample groundwater using boreholes, *Environ. Geol.*, Vol. 32, No. 1, pp 17-22, 1997.

8. Furue, R., Iwatsuki, T., Hama, K., An Appropriate Manner of Hydrochemical Investigation of Groundwater Using Deep Borehole, *J. Jap. Soc. Eng. Geol.,* Vol. 46, No. 4, pp 232-236, 2005.

9. Parkhurst, D.L. and Appelo, C.A.J., User's Guide to PHREEQC (Version 2), Water-Resources Invest Rep. 99-4259, U.S. Geological Survey, Denver, Colorado, 1999.

10. Geotechnical & geoenvironmental software directory, http://www.ggsd.com/ [Accessed: 27-June-2008]

11. Truesdell, A.H. and Jones, B.F., WATEQ, A computer program for calculating chemical equilibria of natural waters, J. Research, U.S. Geological Survey, Vol. 2, pp 233-274, 1974.

12. Matsui, H., Niizato, T., Yamaguchi, T., Horonobe Underground Research Laboratory Project Investigation Report for the 2005 Fiscal Year, Japan Atomic Energy Agency, 2006.

13. Hokari, T., Estimation of In-situ Deep Groundwater Conditions Based upon Chemical Equilibrium Analysis, Tech. Research Report Inst.Technol. Shimizu Corp., Vol. 87, pp 77-86, 2010.

14. Japanese Standards Association (JSA), Water quality – Determination of electrical conductivity, JIS K 0400-13-10, Japanese Industrial Standards, JSA, 1999.

15. Hiraga, N. and Ishii, E., Mineral and chemical composition of rock core and surface gas composition in Horonobe Underground Research Laboratory Project (Phase 1), JAEA-Data/Code 2007-022, JAEA, 2008.

16. Langmuir, D., Aqueous environmental geochemistry, Prentice Hall Inc., Upper Saddle River, New Jersey, 1997.

17. Stumm, W., and Morgan, J.J., Aquatic chemistry: chemical equilibria and rates in natural waters -3rd ed., A Wiley-interscience publication, John Wiley & Sons, Inc., 1996.

18. Kunimaru, T., Shibano, K., Kurigami, H., Tomura, G., Hara, M., Yamamoto, H., Analysis of Ground Water from Boreholes, River Water and Precipitation in the Underground Research Laboratory Project, JAEA-Data/Code 2007-015, JAEA, 2007.

19. Hokari, T. and Kunimaru, T., Estimation of distributions of the in-situ groundwater physic-chemical parameters, *Proceedings of the 2008 Autumn Meeting of Japanese Association of Groundwater Hydrology*, JAGH, Nov. 20-22, 2008, pp 294-299.

20. Masuda, S., Umeki, H., Shimizu, K., Miyahara, K., Naitoh, M., Hasegawa, H., Iwasa, K., The draft second progress report on research and development of HLW disposal in Japan – H12 project to establish technical basis for HLW disposal in Japan – Supporting report 1 – Geological environment in Japan, JNC-TN 1400 99-011, JNC, 1999.

21. Seki, Y., Nakajima, T., Kamioka, H., Kanai, Y., Manaka, M., Tsukimura, K, Discharged water from deep wells in the eastern Kanto region, - the relationship between water quality and underground geology, J. Balneological Society of Japan, Vol. 54, No.1, pp 1-24, 2004.

22. Asamori, K., Umeda, K., Ishimaru, T., Komatsu, R., Information of geological features of the Japanese Islands, JNC TN7450 2002-003, Japan Atomic Energy Agency, 2003.

23. Hem, J.D., Study and interpretation of the chemical characteristics of natural water, University Press of the Pacific, Honolulu, Hawaii, 1970.

13

The 'Shoots Barrage': An Indicative Energy Technology Assessment of a Tidal Power Scheme

Geoffrey P. Hammond[*1], *Craig I. Jones*[2], *Rachel Spevack*[1]
[1]Department of Mechanical Engineering, University of Bath, Bath., UK
e-mail: ensgph@bath.ac.uk
[2]Circular Ecology Ltd., Bristol., UK

ABSTRACT

Several tidal power schemes have been proposed for the River Severn Estuary between the South West of England and Wales. An indicative technology assessment has been undertaken in order to evaluate the so-called 'Shoots Barrage' over its foreseen lifespan of 120 years in terms of its cradle-to-site, operation and maintenance requirements. It would be located just upriver of the Severn road crossings in the United Kingdom (UK), involve an estimated cost of £3.2 bn to construct, and could potentially generate around 2.7 TWh per year (or a little under 1% of UK electricity demand). This scheme is favoured by environmental groups, because to its more benign environmental impacts compared with the much larger, Cardiff-Weston scheme. The present analysis suggests that the proposed Shoots Barrage would yield relatively attractive 'figures of merit' in terms of its net energy and carbon emissions, although its financial performance is poorer than alternative power generators.

KEYWORDS

Shoots barrage, Tidal power scheme, Energy analysis, Carbon accounting, Financial investment appraisal, Sustainability.

INTRODUCTION

Background

Electricity generation presently contributes approximately 30% of United Kingdom (UK) carbon dioxide (CO_2) emissions [1, 2], the principal 'greenhouse gas' (GHG) having an atmospheric residence time of about 100 years [3]. This share mainly arises from the use of fossil fuelled (coal and natural gas) power stations. Changes in atmospheric concentrations of GHGs affect the energy balance of the global climate system. Human activities have led to quite dramatic increases since 1950 in the 'basket' of GHGs incorporated in the Kyoto Protocol; concentrations have risen from 330 ppm to about 430 ppm currently [4]. The cause of the observed rise in global average near-surface temperatures over the second half of the 20th Century has been a matter of dispute and controversy. But the most recent (2013) scientific assessment by the *Intergovernmental Panel on Climate Change* (IPCC) states that it is 'extremely likely' that humans are the dominant influence on the observed global warming since the mid-20th Century [4]. The British Government has therefore introduced a tough, legally binding target of reducing the nation's CO_2 emissions overall by 80% by 2050 in comparison to a 1990 baseline [5] in their 2008 *Climate Change Act* [6]. Achieving this

---

* Corresponding author

carbon reduction target will require a challenging transition in Britain's systems for producing, delivering and using energy that is not only low carbon, but also secure and affordable; thus resolving the so-called energy policy 'trilemma' [7].

The River Severn Estuary lies between the South West of England and Wales in the United Kingdom (UK). It experiences the second largest tidal range (~14 m) in the world and, over the years, a large number of private and UK Government studies have looked for ways to harness the tidal power for electricity generation [8]. But the concept of a Severn Barrage has remained at the feasibility stage since the 1920s, due mainly to concerns about economic viability and environmental impact [8, 9]. Nevertheless, with growing concern over anthropogenic climate change and a desire to ensure a secure energy supply as fossil fuels diminish, the UK Government have committed itself to both the carbon reduction target incorporated in its 2008 *Climate Change Act* [6], and to producing at least 15% of its energy from renewable sources by 2020 [10]. A large-scale Severn Barrage tidal power scheme that was operational by 2020 [11] would provide an estimated supply of 4.4% of the total energy demand of the UK [12]. This would be the so-called Cardiff-Weston barrage (Figure 1) that would be constructed between Lavernock Point near the town of Barry (on the south Wales coast) and Brean Down in Somerset (adjacent to Weston-super-Mare). Thus, by exploiting the tidal range in the Severn Estuary, the UK could improve the energy diversity of its supply mix via such a renewable and sustainable source. A tidal power project in the Severn Estuary could therefore make a significant contribution to reducing GHG emissions from the power sector, as well as helping to meet both international and domestic climate change targets [13].

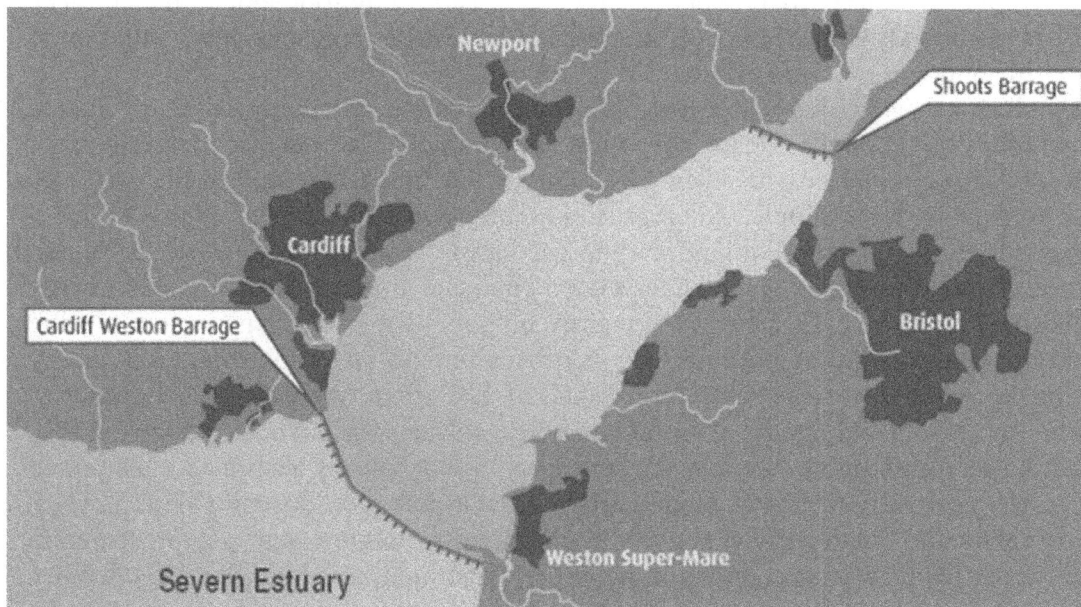

Figure 1. The locations of potential tidal power schemes in the Severn Estuary
(Source: Adapted from the UK *Sustainable Development Commission* [12])

The UK Government's *Department of Energy and Climate Change* (DECC) shortlisted a number of tidal power schemes, including tidal barrages, as well as some alternative, embryonic schemes which would take advantage of the tidal stream. The so-called 'Shoots Barrage' scheme is one of several different tidal barrage possibilities for the Severn Estuary. It would be located upriver of the two present Severn road crossings (see again Figure 1), involve an estimated cost of £3.2 bn (pounds sterling; ISO code: GBP) to construct, and could generate around 2.7 TWh/year (or a little under 1% of current UK electricity demand). The Shoots Barrage is favoured by a number of environmental groups, such as *Friends of the Earth*, due to its lower environmental impact compared, for example, to the much larger Cardiff-Weston scheme.

The issues considered

In October 2010 the new UK Coalition Government announced, following a 2-year cross-government feasibility study of different Severn Estuary tidal barrage and lagoon schemes [13] that it could not see a strategic case for public investment in a Severn tidal power scheme in the immediate term, though private sector groups would continue to investigate the potential. The costs and risks for the British taxpayer and energy consumer were regarded as being too high in the current financial situation, i.e., the post-2008 economic recession. However, it wished to keep the tidal barrage option open for future consideration. The decision not to rule out a scheme in the longer-term recognises its significance as a large-scale UK energy resource. There were half a dozen substantive responses to this announcement from organisations like the *Bristol Port Company*, the *Countryside Council for Wales*, the *Environment Agency*, WWF UK, and the consulting engineers *Parsons Brinckerhoff*. They argued that work should start now in order to:

- Address the significant uncertainties and data gaps;
- Monitor the detailed baseline of distribution of animal species and habitats;
- Study fish behavior and movement in the estuary;
- Assess measures to prevent or reduce possible environmental impacts. The present study of the Shoots Barrage scheme thereby represents a contribution to this ongoing research effort.

An indicative technology assessment has been conducted comprising a detailed investigation into the cradle-to-site, operation and maintenance energy consumption for the Shoots Barrage tidal power scheme (Figure 1). An 'integrated approach' was used (similar to that of, for example, Allen *et al.* [14]) to assess the impact of this scheme, employing both energy analysis and carbon accounting applied on a 'whole systems' basis from 'cradle-to-grave', alongside related financial investment appraisal. Energy analysis (EA) required estimates of the energy outputs of the power generators during use, and the energy requirements for their construction and operation. The total energy output of the scheme over its foreseen lifespan of 120 years was estimated in order to determine the associated energy gain ratios (EGR) and energy payback periods (EPP). But carbon footprints have become the 'currency' of debate in a climate-constrained world. They represent the amount of carbon (or carbon dioxide equivalent [CO_{2e}]) emissions associated with a given activity or community, and are generally presented in terms of units of mass or weight (kilograms per functional unit [e.g., kg CO_{2e}/kWh]). Embodied energy and carbon appropriate to the various power generators specified in the current work were determined using the '*Inventory of Carbon and Energy*' (ICE) [developed at the University of Bath (Hammond and Jones [15, 16])]. 'Embodied energy' is here defined as the total primary energy consumed from direct and indirect processes associated with power production and within the boundary of 'cradle-to-gate' [16]. This includes all activities from material extraction (quarrying/mining), manufacturing, transportation and right through to fabrication processes until the power plant is constructed for operational use. Similarly, 'embodied carbon' is the sum of fuel-related carbon emissions (i.e., embodied energy which is combusted, but not the feedstock energy which is retained within materials) and process-related carbon emissions [16].

The present contribution is part of an ongoing research effort aimed at evaluating and optimising the performance of alternative sustainable, centralised and distributed energy systems for the UK [14, 17-19] in the context of transition pathways to a low carbon future for the UK [7, 20]. Here the 'Shoots Barrage' tidal power scheme has been evaluated using various appraisal techniques to determine its net energy output, carbon footprint, and financial investment issues. This study is 'indicative' in the sense of being a simplified evaluation and illustration of the performance of this tidal power scheme in the light of imperfect information. Thus, the uncertainties involved are quite large, because of the rough estimates available at the concept design stage of the proposal.

THE SHOOTS BARRAGE

Overview of the scheme

The Shoots scheme is a proposed 1.05 GW barrage located upriver of the Second Severn Crossing (Figure 1); its position would coincide with the highest tidal range in the Severn. The site was first investigated by the first Severn Barrage Committee (1925-33) under Lord Brabazon at which time it was referred to as 'The English Stones Scheme'. The *Severn Tidal Power Group* (STPG) in 1986 [21] studied this barrage in detail alongside the Cardiff-Weston barrage. They raised concerns regarding the rate at which sediment could build up in the basin. The latest proposed Shoots Barrage scheme is outlined in Table 1 as more recently examined by the UK Government's independent *Sustainable Development Commission* (SDC) [12]; established by the then Labour Government in 2000 (although the subsequent Coalition Government withdrew funding after coming to power in 2010, and the SDC had to close in March 2011). This barrage scheme is potentially able to generate some 2.75 TWh per year using 'Straflo', or rim generator, turbines [8] operating via solely ebb generation (see the schematic representation in Figure 2). Thus, the incoming flow is allowed to pass through the barrage sluice gates, where the water is trapped behind the barrage at high-tide by closing the sluice gates [8]. The head of water then drives water back through the turbines on the outgoing or 'ebb' tide in order to generate power. A single navigation lock was included in the proposed SDC scheme [12] which is able to handle ships up to 25,000 deadweight tonnes (dwt), allowing the English port upstream of this proposal at Sharpness in Gloucestershire to remain fully functional. Sharpness handles approximately 400 vessels per year. The Shoots Barrage scheme, which was analysed by the consulting engineers Parsons Brinckerhoff, abates the concerns of the STPG report [21] relating to the rate of silting in the basin through the use of high-level sluice gates [12]. These gates would close the turbine during the flooding of the basin and exclude the lower part of the flow, which is more sediment rich. It should be noted, however, that Parsons Brinckerhoff recommended that further analysis should take place at the next design phase to corroborate their findings.

Table 1. Outline of the proposed shoots tidal barrage scheme
(Source: The UK *Sustainable Development Commission* [12])

Length of embankments	4.1 km
Generating capacity	1.05 GW (1,050 MW)
Annual average electricity output	2.75 TWh
Number of turbines	30
Number of sluice openings	42
Ship lock size	225 m × 37.5 m

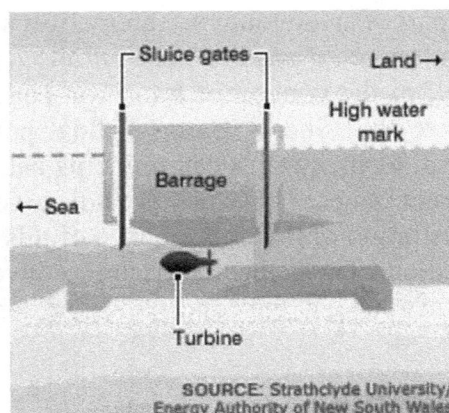

Figure 2. The 'Shoots Barrage': based on ebb generation using 'straflo' turbines

Construction methods and costs

The construction method for the Shoots Barrage scheme involved the towing of caissons out to site and then they were sunk into place. Similarly, the navigation lock would consist of a single steel caisson, which would be fully fitted out before being placed onsite [12]. This site was assumed able to provide better foundations for the barrage than the Cardiff-Weston site, due to inter-tidal rock outcrops. These enable rather simpler construction arrangements. Parsons Brinckerhoff/Black & Veatch [11] estimated the need for a 4 year design and planning phase, followed by a construction period of 5 years (2014-2019). The operation and maintenance phase of the barrage would then last over the period 2019-2140, when decommissioning would commence. The Parsons Brinckerhoff/Black & Veatch options analysis report for DECC [11] assumed a constant annual expenditure for the pre-construction period, as well as a constant expenditure during the construction period. A more realistic breakdown of the latter expenditure was adopted for the present study based on the profile given in the 1986 STPG report [21]. No inclusion of the cost for public road construction or of a rail link was included in either the DECC-sponsored report [11] or in this present study.

Maintenance, operation, and decommissioning

The methods used to calculate the maintenance and operational costs vary within earlier studies. That by Parsons Brinckerhoff/Black & Veatch for DECC [11] presents a 'worst case' having 70% of the present value for the supply, installation and commissioning costs of mechanical and electrical equipment being incurred every 40 years. The cost of the turbine generators was estimated at £5,841 mln [11]; equivalent to £817 mln per annum over the 5 year maintenance period. In order to estimate the corresponding energy requirements and carbon emissions during the maintenance period; 70% of the total emissions for the mechanical and electrical (M&E) equipment was, therefore, adopted. DECC took a value of 1.75% of the total construction cost as the annual maintenance cost of the project (equal to £314 mln). This figure was assumed to remain constant even when the barrage is running at only 25% total output during the maintenance years. The costs associated with the decommissioning of the barrage have been excluded from this study – a practice that is in-line with all earlier studies [11-13, 21]. The decision to disregard this, potentially significant, item is due to the long design life of the barrage of 120 years. Over the past 120 years attitudes to decommissioning and recycling have changed significantly, as have decommissioning/recycling methods, and it is therefore not possible to predict how future generations would dispose of a tidal barrage.

ENERGY ANALYSIS

Methodology

In order to determine the primary energy inputs needed to produce a given artefact or service, it is necessary to trace the flow of energy through the relevant industrial sector [14, 15, 17, 19, 20, 22]. This is based on the *First Law of Thermodynamics* (the principle of conservation of energy) or the notion of an energy balance applied to the system. The system boundary should strictly encompass the energy resource in the ground [23-25] (known as the 'cradle' - for example, oil in the well or coal at the mine). In the present analysis the downstream boundary is known as the 'site' (hence, 'cradle-to-site' [16, 20]), or national electricity network (operated by the 'transmission network operators' [TNOs] and 'distribution network operators' [DNOs]). Consequently, it effectively accounts for all UK power sector primary energy use (and associated emissions). Energy analysis yields the whole-life or *'Gross Energy Requirement'* (GER) of the product or service system [15, 23-25]. Thus, the sum of all the outputs from this system multiplied by their individual energy requirements must be equal to the sum of inputs multiplied by their individual requirements. The process consequently implies the identification of feedback loops, such as the indirect or 'embodied' energy requirements for materials and capital inputs. Several differing methods of EA have been developed (see Figure 3), the most significant being statistical analysis, Input-Output (I-O) analysis, process analysis (or energy 'flow charting'), and hybrid analysis [15, 23-25].

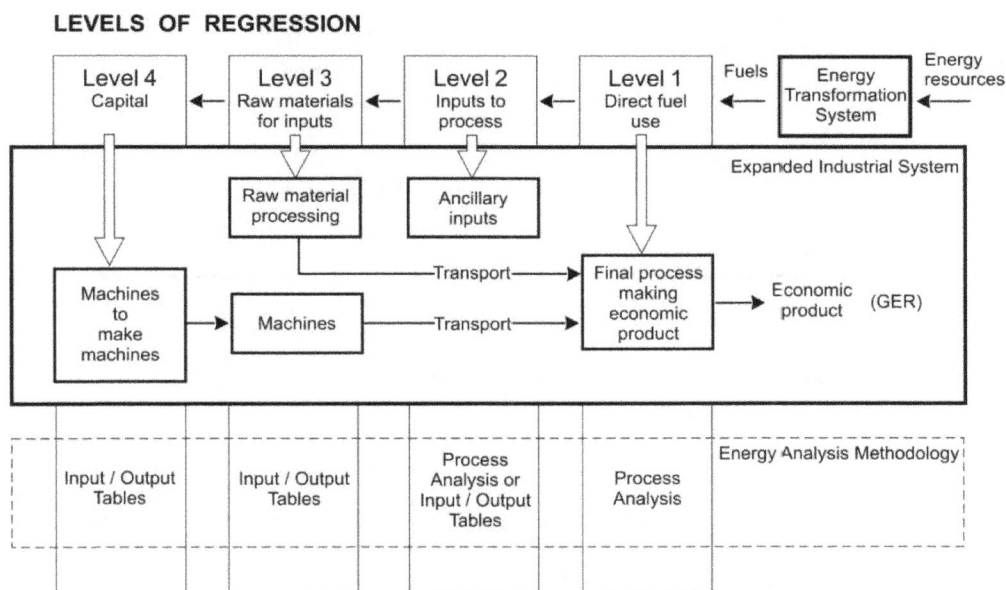

Figure 3. Schematic representation of the energy analysis process
(Source: Allen *et al.* [14]; adapted from Slesser [25])

Application of energy analysis to the shoots barrage

Energy analysis, as indicated above, is an established method of tracing the flow of energy through a system [14, 15, 17, 19, 20, 22-25], and can be readily applied to large-scale civil engineering projects. The present analysis has been conducted in order to assess and compare the envisaged energy benefits of the proposed Shoots Barrage scheme as a more benign option for the generation of electricity than those from fossil fuels. The methods used to carry out an EA mainly stem from studies completed in the 1970s (e.g., [23-25]). They can account for, and hence suggest ways to reduce, the energy consumed and expended over the lifetime of the system under consideration. This includes the embodied energy of the raw materials, transportation, construction, maintenance, operation, and decommission. The stages investigated in process energy analysis employed here are illustrated in Figure 4. It can be seen that there was limited data available some processes, and where simplifying assumptions needed to be made. Other processes had to be excluded from the study because of the unavailability of suitable data. These generally had an insignificant impact on the life-cycle energy requirements of the barrage.

Component fabrication, which has been excluded from the present study, refers specifically to the fabrication of items such as the turbine generators and the caissons. No accurate data was available to account for the direct energy required to manufacture the turbines. However, the raw materials required in the manufacture of the turbine (which accounts for the bulk of the energy requirements), and the transportation from the manufacturer to the barrage site were estimated. Likewise, no data exists for the energy required to fabricate items such as the ship locks and caissons, and this was again excluded from the energy analysis.

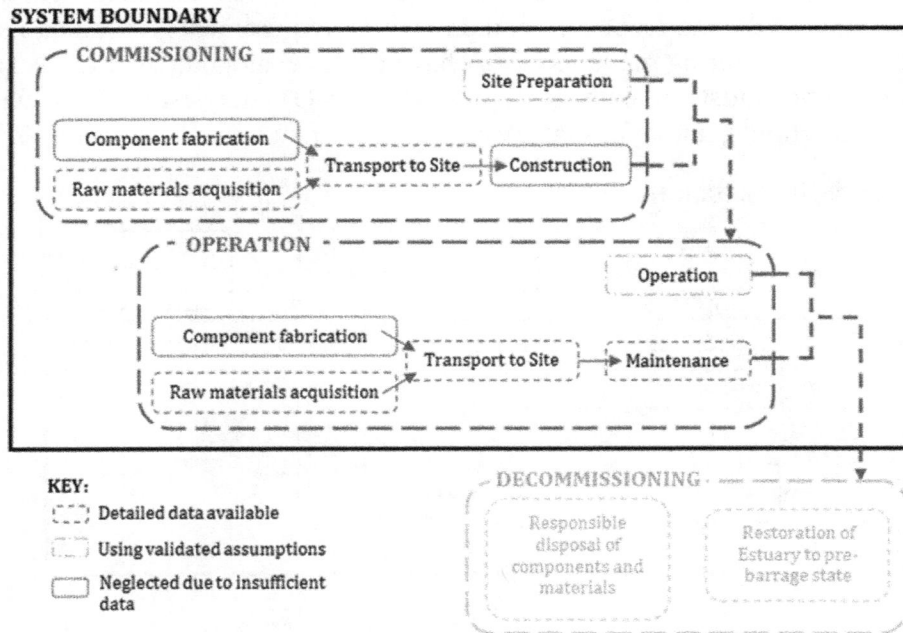

Figure 4. System boundary for tidal power energy analysis and carbon accounting

Calculation of the Energy Gain Ratio (EGR)

The energy gain ratio (EGR) divides the useful energy produced by the barrage over its lifespan by the total energy consumed from cradle-to-grave [25]. The net energy produced is the total net electricity generation - converted from Watt-hours to Joules for consistency [26]:

$$\text{Energy Gain Ratio} = \frac{E_{n,L}}{E_{mat,L} + E_{con,L} + E_{op,L} + E_{dec,L}} \tag{1}$$

where $E_{n,L}$ is the net energy produced over the lifetime (L) of the barrage; $E_{mat,L}$ is the total energy invested in materials; $E_{con,L}$ is the total energy invested in construction; $E_{op,L}$ is the energy required to operate the plant over its lifetime, and $E_{dec,L}$ is the energy required to decommission the barrage at the end of its life. The energy required to decommission the barrage (the 'grave') is outside of the system boundary for this energy analysis (see again Figure 4 above). This is in-line with all earlier studies of the Severn Barrage schemes [11-13, 27]. It has only been possible to partially examine the energy required to construct the Shoots Barrage, because of the rough estimates available at the pre-detailed design stage of the proposal.

Calculation of the Energy Payback Period (EPP)

The energy payback period (EPP) represents the period (the number of years) that a renewable energy (RE) device must operate before it has captured and delivered as much primary energy as has been used to construct the RE technology [24, 25]. The values calculated here were obtained on the basis of a 'static' energy analysis approach [25]. The number of years at which the electricity generated by the barrage equals the primary energy invested in the barrage is the EPP.

The opportunity cost convention

In the discipline of economics the notion of 'opportunity cost' relates the financial opportunity or return that is foregone when an investment is made in one project (the opportunity) in contrast to an alternative [24]. Thus, the equivalent convention in EA concerns the energy foregone in order to provide energy via another conversion process. In the power sector, fossil fuels (thermal or primary energy) are typically invested in constructing conventional plants rather than in low carbon alternatives, such as nuclear power or various renewable energy technologies. In the tidal power case, the opportunity cost (or 'opportunity energy requirement') will therefore represent the primary energy foregone during the construction of a barrage that is required to generate electricity over its lifetime [24, 25]. In order to calculate the opportunity cost (OC) in the present study, the weighted average efficiency of the electricity sector (η) was taken as 38.5% (see, for example, Hammond [3]). The OC equivalent of the standard EGR and EPP above is then obtained by:

- Dividing the former by η;
- By multiplying the latter by η and adding the initial construction period (in years).

Assumptions and approximations

Construction materials: Gate-to-site. The cradle-to-site energy requirements include the raw material extraction, processing and transportation to the construction site [15]. These reflect the 'embodied energy' associated with these activities, i.e., the total primary energy consumed from direct and indirect processes associated with power production and within the defined cradle-to-site boundary as indicated in Figure 4 above [16]. In the present analysis, the transportation was examined separately on such a basis. Thus, information relating to the quantity of each raw material required in construction was taken from a 2007 report compiled by Black & Veatch (a global engineering, consulting, and construction company) for the SDC [12]. The University of Bath's ICE database (v2.0) [15, 16] was then used to determine the embodied energy relating to the raw materials. This database provides a range of embodied energy figures associated with material component, along with an indicated of the prevailing scatter in the data. Woollcombe-Adams et al. [26] estimated the carbon emissions for the Cardiff-Weston Severn Barrage tidal barrage scheme. They assumed distances that raw materials would have to be transported and, combined with the mass of the material, determined the total carbon emissions. However, the basis for these assumptions is unclear. Therefore, in order to determine transport distances for the present study, suitable quarries or manufacturers in closest proximity to the barrage site were identified. Having located these sources and the quantities of raw materials required, the primary energy consumption was calculated using data taken from a 2008 report by the *Institut fur Energie und Umweltforschung* (IFEU) [28]. The IFEU report provides coefficients relating the primary energy consumption of various modes of transport to the weight of the material transported and the distance travelled.

Construction. Little information is available relating to the exact work requirements to construct the barrage. A decision was therefore made to neglect the construction of components, such as the turbine generator and caissons, in this study. This amounts to an EA terminated at Level 3 Regression as indicated in Figure 3. However, data for other items (such as dredging and the towing energy) was accounted for here. Roberts [29] provides information on caisson building for various large tidal barrages. It has been assumed for the current purposes that each casting yard consumed 1.75×10^6 GJ during construction.

Dredging. In order to calculate the energy consumed by dredging, the estimates used by Roberts [29] were again adopted to determine the energy required to extract this material from a quarry. This value has been verified by comparison with the data in the ICE database (v2.0) [16]. In the case of the Shoots barrage, the embodied energy is significantly lower using the dredging figure given by Roberts [29] than that in the ICE database [15]. It is believed that this is due to the difference in the geology type at the two sites considered by Roberts and the Shoots site. Owing to the far smaller initial investment energy of the Shoots scheme, and the lower energy delivered following commissioning, the value used had a significant impact on the final energy gain ratio and energy payback periods. It was therefore decided that the value derived using the ICE data (v2.0) [16] related to a rock-based foundations was most appropriate to the Shoots scheme.

Towing energy. Roberts [29] assumes what he termed 'towing out' energy gave rise to energy consumption of 54×10^6 MJ/caisson. This figure has been compared to the energy consumed to transport each caisson a distance of 100 km; to represent an approximate distance between a barrage and fabrication yard. Here the *EcoTransIt* database [28] was used to estimate towing for the purposes of energy analysis. In the case of the Shoots Barrage, it is presently uncertain as to where the caissons would actually be constructed. Only the float-out weights of the Cardiff-Weston caissons are presently known, each at 126,000 tonnes [12] for the heaviest of the caissons. All caissons have been assumed to be of the same mass for the Cardiff-Weston and Shoots barrages; thereby representing a worst case. This suggests that the towing energy had a value of 9.1×10^6 MJ/caisson. Thus, the data provided by Roberts [29] above appears to be pessimistic. To account for the towing energy required to install the ship locks, the segments of the locks were approximated to the same float out weight as the caissons. The Shoots scheme is presumed to be composed of just one segment for its ship lock.

Operation and maintenance allowances. Roberts [29] adopted a value for the energy intensity equivalent to 5.28 MJ/£ (2010) to account for the annual operational cost of the barrage. This represents about 1.75% of the total capital cost, in accordance with data more recently provided by the DECC [11] for options analysis of the development of tidal power in the Severn Estuary. A similar share of the total embodied energy from construction per year of operation was adopted for the present energy analysis. Maintenance was assumed to be required every 40 years, and hence there would be two 2-year maintenance periods for the Shoots scheme over its total envisaged lifespan of 120 years. An assumption was made that 70% of embodied energy related to the manufacture of turbines, their transport and installation (in line with the financial analyses published by the DECC [11], where they made an allowance of 70% for the mechanical and electrical [M&E] equipment costs). This resulted in 2×10^6 GJ over a 120 year lifespan for the Shoots barrage.

CARBON ACCOUNTING

Methodology

It is widely recognised that in order to evaluate the environmental consequences of a product or activity the impact resulting from each stage of its life-cycle must be considered [22]. This has led to the development of a range of analytical techniques that now come under the 'umbrella' of environmental life-cycle assessment (LCA). One of the antecedents of this approach was energy analysis of the type described above. In a full LCA study, the energy and materials used, and pollutants or wastes released into the environment as a consequence of a product or activity are quantified over the whole life-cycle; 'from cradle-to-grave' [30, 31]. The methodology of LCA follows closely that developed for energy analysis [14, 20, 22, 25], but evaluates all the environmental burdens associated with a product or process over its whole life-cycle. This requires the determination of a balance or budget for the raw materials and pollutant emissions (outputs) emanating from the system. Energy is treated concurrently, thereby obviating the need for a separate EA [22]. LCA is often geographically diverse; that is, the material inputs to a product may be drawn from any continent or geo-political region of the world [15]. But, as previously argued, carbon footprints have become the 'currency' of debate in a climate-constrained world. Consequently, the emphasis in the present study was on CO_2 emissions, rather than the wider set of environmental burdens [14, 17, 19, 20, 27]. An emissions coefficient (in gCO_2/kWh_e) for the Shoots barrage scheme was calculated using as expression derived by White and Kulcinski [26]:

$$\frac{kg \times CO_2}{kWh} = \frac{\Sigma_i \left(\frac{kg \times CO_2}{kg \times M_i} \right) \times kgM_i}{E_{n,L}} \qquad (2)$$

where $E_{n,L}$ is the net electrical energy produced over the lifetime of the barrage, L; $kgCO_2$. M_i is the kg of CO_2 emitted per kg of material i produce; kgM_i is the quantity of material i needed to constructed and/or operate the barrage. The same methods used to calculate the embodied energy [15] and other primary energy requirements have been applied to the carbon analysis.

Assumptions and approximations

In order to calculate the cradle-to-site CO_2 emissions a similar approach was taken to that employed for the energy analysis described above (see Figure 4). The University of Bath's ICE database (v2.0) [16]; was again used to determine the cradle-to-gate CO_2 emissions associated with raw materials employed found for the EA. The gate-to-site emissions were then calculated per tonne of material per km travelled. However, the energy consumed to construct the Shoots Barrage has been neglected due to a lack of available data. No specific data exists relating to the carbon emissions generated during dredging. The emissions released to quarry the materials were therefore extracted again from the ICE database [15]. This approach provided comparable results to that employed in the EA. Carbon emissions during towing out have been ignored due again to insufficient data being available. The method used for the energy analysis (employing the EcoTransIt database [28]) did not adequately represent the 'towing out' energy for the caissons, and hence cannot reliably be used to estimate carbon emissions. The carbon emissions generated during the annual operation of the Shoots Barrage, as well as the maintenance periods every 40 years, were estimated by adopting the same approximations as described for the EA above. Maintenance has been equated to 70% of the total M&E equipment carbon emissions, producing 0.13×106 tonnes CO_2 for the barrage. Annual operational carbon emissions have been taken as 1.75% of the total emissions released during construction; this converts to 1.77×106 tonnes CO_2 for the Shoots scheme based on a 120 year lifespan. But CO_2 emissions released during the projected decommissioning phase were again not been accounted for.

FINANCIAL INVESTMENT APPRAISAL

Methodology

Background. Economic appraisal evaluates the costs and benefits of any project, programme, or technology in terms of outlays and receipts accrued by a private entity (household, firm, etc.) as measured through market prices [32]. Financial appraisal is used by the private sector and omits so-called environmental 'externalities'. In contrast, *economic cost-benefit analysis* (CBA) is applied to take a society-wide perspective, with a whole systems view of the costs and benefits [14, 22]. It accounts for private and social, direct and indirect, tangible and intangible elements; regardless as to which they accrue and whether or not they are accounted for in purely financial terms [32]. Allen *et al*. [14] applied both financial appraisal and CBA to evaluate a number of micro-generators, whereas Hammond *et al*. [19] more recently used them to evaluate a building-integrated solar photovoltaic (PV) array. A further distinction between *financial appraisal* and *CBA* is in the use of the discount rate to value benefits and costs occurring in the future [14, 19, 22]. Financial appraisal uses the market rate of interest (net of inflation) as a lower bound, and therefore indicates the real return that would be earned on a private sector investment.

Capital expenditure and the breakdown of annual costs. The capital expenditure associated with a Severn Estuary tidal barrage project was taken from a 2008 study sponsored by DECC [11]. The report on this study provides detailed cost estimates for the Cardiff-Weston scheme in terms of construction, electricity generation, and operational costs. Scaled figures were applied in the present work to the Shoots scheme. The STPG [21] report described in detail a capital cost breakdown over 6 year pre-construction period. This was compressed to fit the construction period of 5 year period envisaged by DECC [11]. The maintenance costs were again approximated at 70% of the M&E generating plant every 40 years, with a maintenance period of 2 years for the Shoots barrage. The annual operation cost of the barrage was taken as 1.75% of the total construction cost for the scheme. This is in line with the estimates made by DECC [11], although they state that in the case of the Cardiff-Weston barrage they estimated an annual cost of just 1.25%. The cost of decommissioning the Shoots barrage has again not been accounted for in the present study.

Compensatory habitat. An additional allowance for compensatory habitat has only been included in the most recent studies. For the Shoots scheme, depending on how the total scheme would be funded, there is a potential for this to be provided by the private sector. The impact of the Shoots on the Severn Estuary (see Figure 1 above) is much less significant than for the larger Cardiff-Weston scheme, and therefore the best financial case assumed no compensatory habitat was required. These compensatory habitat costs for the Shoots barrage ranged from £0.32 bn to £0.96 bn.

Discounted cash flow. The *Levelised Unit Electricity Cost* (LUEC) is typically employed to compare the economic performance of different power generators. This is the price at which electricity must be sold in order to recover all costs incurred during generation. The net present value (NPV) of the sum of the capital cost, maintenance and operational costs and, potentially, decommissioning is calculated over the life of the project, along with the NPV of the total electricity generated. This yields the LUEC in *pence per kilo-watt hour* (p/kWh$_e$) for the Shoots barrage, which can then be compared to that for alternative power generators. Consequently, by using this method, different energy options with a variety of lifespans, capital costs, and efficiencies can effectively be compared so that the most cost-effective option can be determined. The discounted cash flow over the life of each project (here assumed to be 120 years) is calculated as follows:

$$\text{Discounted Cash Flow} = \sum_{t=1}^{t=120} \frac{R_t}{(1 + \text{TDR})^t} \qquad (3)$$

where R_t is the net receipts (income less cost); t is the time in years for the total foreseen life of the project, and r is the discount rate. In the case of public sector investments a so-called *Test Discount Rate* (TDR) is utilised. It is typically derived from a comparison with private sector discount rates (or *Weighted Average Cost of Capital* [WACC]). In the UK, HM Treasury [33] recommends that the TDR for projects with durations of less than 30 years should be taken as 3.5%, then falling in line with the profile indicated in Table 2 below.

Table 2. The declining long-term UK test discount rate [33]

Period of years	0-30	31-75	76-125
Discount rate	3.5%	3.0%	2.5%

The results obtained by DECC [11] do not use these TDR values, as they believe that it would not satisfactorily manage all of the risks associated such a project, and will only represent the case of the scheme being entirely funded through the public sector [5]. The LUEC values presented by the DECC [11] employed a discount rate of 8%, which they regard as reflecting the WACC that would enable the project to be financed by the private sector.

RESULTS AND DISCUSSION

Energy analysis

Cradle-to-site analysis. Embodied energy associated with the material requirements for the Shoots Barrage were obtained from the ICE database [15, 16]. The highest contributor in terms of the energy requirements for the barrage was rock (Figure 5), with 97% of this being due to the energy consumed during transportation between the quarry and the barrage site. The assumption made in the gate-to-site analysis was that rock was shipped from the *Glensanda* super quarry in Scotland. The *Severn Barrage Steering Committee* identified this quarry as an alternative option, if it was not possible to source the rock locally in Wales. When investigating items such as crushed aggregate in this study it was found that shipping items from this Scottish quarry was more energy efficient owing to the poor freight train connections in Wales, along with the high energy requirement for moving freight by road. *Glensanda* has its own shipping port, and hence no movement of goods by road occurs.

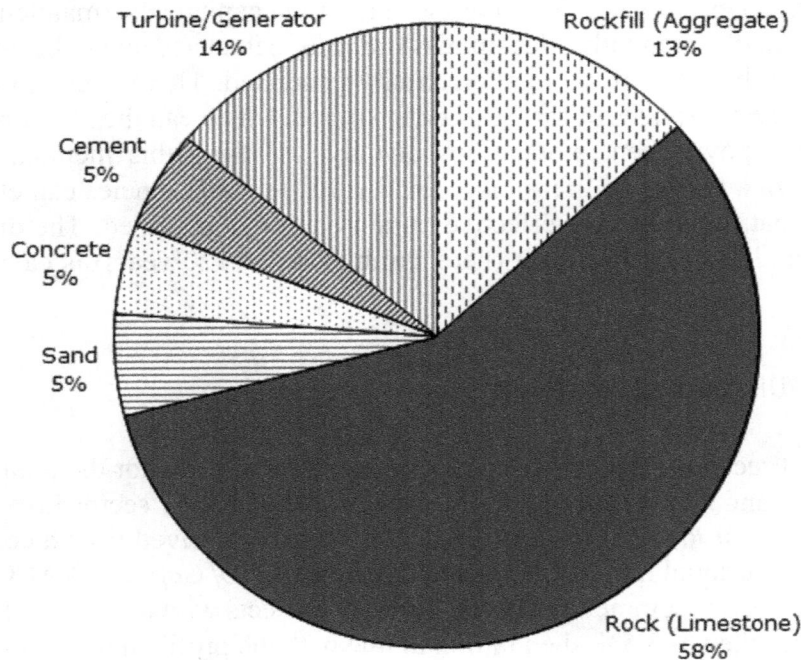

Figure 5. Cradle-to-site energy analysis: Shoots barrage material inputs

<u>Total energy requirements.</u> Energy requirements for commissioning and operation of the Shoots Barrage were analysed. The major element of the total energy consumed was found to arise due to operational energy requirements for the barrage. This suggests that the assumptions made about the commissioning of the scheme have a relatively minor influence on the overall energy requirements. The energy required to fabricate the individual components and the barrage itself was therefore neglected in the present study.

<u>Energy Gain Ratio (EGR).</u> Table 3 displays the EGR calculated using assumptions outlined above. The final energy gain ratios have been put into context by comparing them to other electricity generation plants. It has been possible to recalculate the EGRs for conventional nuclear and coal power plants so that they do not include plant construction or decommissioning energy requirements. These figures can then be more easily compared to the present ones (in Table 3). The EGR for the Shoots Barrage scheme is approximately double that of the coal plant investigated by White *et al.* [26]. The Shoots Barrage gives rise to a slightly better EGR than nuclear (fission) power stations. The EGR for wind, despite taking into account the energy consumed during construction and decommissioning, is higher than those of the Shoots Barrage. An average value has

been taken for the results of the above cradle-to-site analysis, and the EGRs were calculated at three different possible lifespans - from a full lifespan of 120 years down to 40 years [which is slightly less than the 44 year lifespan of the La Rance barrage (located on the estuary of the Rance River in Brittany, France); the longest proven lifespan for this technology to date]. The Shoots Barrage EGR fell to 19.44 for an 80 year life, and 14.00 for one of 40 years.

Table 3. Estimated 'Energy Gain Ratios' (EGRs) for Alternative Power Generators

Scheme	EGR	Lifespan [years]
Shoots tidal barrage	22.31	120
Coal-fired power plant	10.8 (10.8[*])	40
Nuclear power station	17.8 (16.4[*])	40
Wind turbine (without storage)	(23[*])	25

NB: Data for alternative power generators taken from White and Kulcinski [26]. Numbers with an asterisk ([*]) indicates EGRs with the inclusion of plant construction and decommissioning.

The opportunity cost convention. This convention was been applied to the EGR estimated for a 120 year (default) lifespan. It then rose to be between 22.31:1 and 57.9:1. In terms of energy, this obviously makes the Shoots Barrage scheme a highly attractive power generation option on a like-for-like basis.

Energy Payback Period (EPP). The EPP indicates the time taken, in months following first operation, for the amount of energy generated by the Shoots Barrage to equal the energy consumed during commissioning and operation up to that moment in time. It should again be noted that not all of the energy consumed during this phase could be accounted for in this study, and hence it should be assumed that the current figures are optimistic. The EPP for the Shoots scheme is 50 months from first commissioning; assuming just a 50% power generation capacity in year one. Taking a construction period of 5 years, this equates to a total energy payback of 9.16 years. This is slightly longer than that of the much larger Cardiff-Weston scheme, but this difference is not significant over a lifespan of 120 years. By applying 'opportunity cost' convention, the EPP for the Shoots barrage was only 6.60 years, which indicates a strong case (in energy terms) for the implementation of such a scheme.

Carbon accounting

Cradle-to-site emissions. Two earlier cradle-to-gate studies of the carbon emissions have previously been completed by the SDC [12] and by Woollcombe-Adams et al. [27]; the data obtained during the present study was compared to these two previous sources. These employed an embodied energy value for rock that was slightly higher than that adopted here, but the material type was confirmed by a member of the Severn Barrage Steering Committee. The carbon coefficient for extracting the material from the quarry is taken in the present study from a more recent version from the ICE database (v2.0) [16]. Data from previous studies slightly underestimate the GHG emissions from that raw material. The extraction of the required quantities of cement produces the greatest quantity of carbon emissions. Only the SDC study [12] - undertaken for them by Black & Veatch - has published detailed results, and these do not compare well with those from present study (Figure 6). Variances are likely to be mainly due to different assumptions about the material content of the barrage, and to a lesser extent to the fact that Black & Veatch used an older version of the ICE database (v1.5).

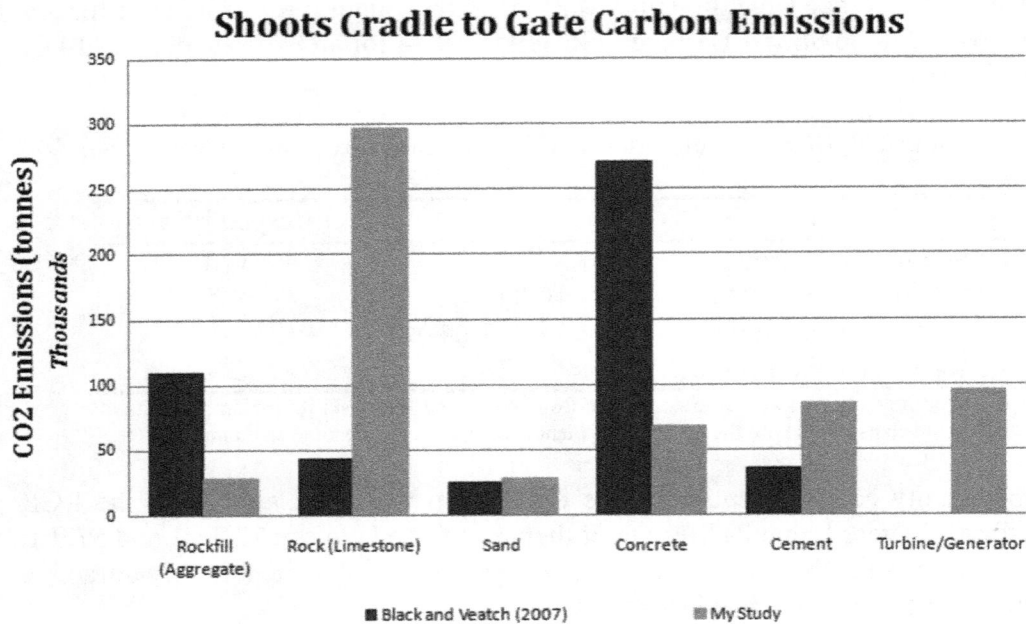

Figure 6. Shoots barrage: Cradle-to-gate carbon emissions
(Sources: Black & Veatch for the UK *Sustainable Development Commission* [12]; 'My Study' represents the present work)

Total carbon emissions. The highest carbon emissions occur during the operational phase of the Shoots Barrage (around two thirds), as in the case of the energy requirements calculated above. The assumption of 1.75% of the total emissions during construction has been taken from the financial model first made by the DECC (11). It is far higher than in earlier report by the STPG (21). However, the proportion of 'on-site' carbon emissions is smaller than that found in the energy analysis above. This was in part due to items, such as towing energy requirements, being ignored in the present analysis. The total carbon emissions over the assumed 120 lifespan were some 2.75 MtCO$_2$.

Carbon dioxide emissions per unit of electricity. Estimates of the carbon dioxide emissions per unit of electricity generated (in gCO$_2$/kWh$_e$) were made assuming a lifespan of 120 years. The range of CO$_2$ emissions was calculated by applying the range of data obtained at the cradle-to-gate phase. The overall results show that the Shoots Barrage scheme would emit about 8.0 gCO$_2$/kWh$_e$. That is attractive in terms of a low level of carbon emissions. But it should be noted that not all of the sources of emissions could be accounted here, and the actual results would consequently be slightly higher in reality.

Financial investment appraisal

Baseline LUEC. This analysis provides a baseline cost for the Shoots Barrage of 10.83 p/kWh$_e$ (derived for a 120 year lifespan, using the STPG cost breakdown): see Figure 7. The results obtained using the DECC [11] investment appraisal approach with constant capital expenditure over the pre-construction and construction periods are slightly higher than those using the more detailed breakdown derived from the STPG [21], although this does not represent a significant difference. The ultimate Levelised Unit Electricity Cost (LUEC) variation over a 120 year lifespan using the declining TDR (see Table 2; advocated by HM Treasury) produced a value of 4.72 p/kWh$_e$ for the STPG breakdown and 4.67 p/kWh$_e$ for that the DECC study. When applying a TDR of 8%, this range rose to 10.83 p/kWh$_e$ for the STPG method (the baseline case as indicated above) and 10.42 p/kWh$_e$ from the DECC study. Differences in the LUEC determined by using a more detailed cost breakdown by the STPG [21] in comparison with than the constant expenditure model of DECC [11] produced a difference of only 0.41 p/kWh at a discount rate of 8% over a 120 year lifespan. The SDC-sponsored study [12] by Black & Veatch indicated LUEC values for the Shoots tidal barrage of 3.29 p/kWh$_e$ (with a range of 2.96-3.62 p/kWh$_e$) for a social discount rate of 3.5% or 6.8 p/kWh$_e$ (with a range of 6.08-7.52 p/kWh$_e$) for an investor discount rate of 8%.

Shoots LUEC Range of Present Study Compared to DECC

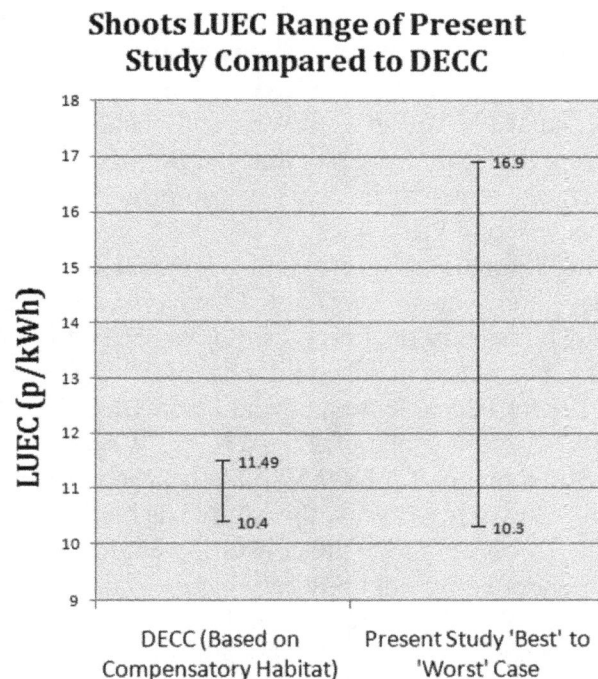

Figure 7. Shoots barrage: Comparison of the 'best' to 'worst' case ranges of LUEC studied here to that of DECC [11]

CONCLUSIONS

Several tidal power schemes have been proposed for the River Severn Estuary between the South West of England and Wales. Here the so-called Shoots Barrage scheme has been evaluated (see Figures 1 and 2) using various appraisal techniques to determine its net energy output, carbon footprint, and financial investment issues. It would located near the Severn road crossings in the United Kingdom (UK), involve an estimated to cost £3.2 bn to construct, and could generate around 2.7 TWh/yr [or just about 0.7% of UK electricity supply]. An energy analysis was conducted comprising a detailed investigation into the cradle-to-site, operation and maintenance energy consumption for the two schemes. The total energy output of the scheme over its foreseen lifespan of 120 years was calculated in order to determine the associated *energy gain ratios* (EGR) and *energy payback periods* (EPP). The former was found to vary from

19.2:1 to 23.8:1 (see Table 3), whilst the latter was estimated to be about 9.16 years. On an 'opportunity cost' basis the EGR rose to be between 22.3:1 and 57.9:1 with an EPP of about 6.6 years. Overall, the present analysis suggests that the Shoots scheme has relatively attractive 'figure of merit' in energy terms.

The above system boundary (see also Figure 4) was then applied for carbon accounting, and this yielded a 'footprint' of about 8.0 gCO_2/kWh_e. In both the energy and carbon analyses, the operational requirements/emissions of the Shoots Barrage were found to have the most significant influence on the final results (accounting for around two thirds of the emissions). It was not possible to include all of the energy requirements associated with the scheme or the sources of carbon emissions from the project, such as those emanating from the manufacturing of the turbines and caissons. However, they are unlikely to have a significant impact on the energy/carbon indicators estimated here. The Shoots Barrage is favoured by environmental groups, such as Friends of the Earth, due to its less severe environmental impacts than the larger, Cardiff-Weston scheme. Work sponsored under the auspices of the IPCC [34] indicates the carbon intensity of alternative power generators: coal (without carbon capture and storage [CCS]) ~1000 gCO_2/kWh_e; combined cycle gas turbines (CCGT; without CCS) - 443 gCO_2/kWh_e; nuclear - 66 gCO_2/kWh_e; solar PV - 32 gCO_2/kWh_e; and onshore wind - 10 gCO_2/kWh_e. Again, the present analysis therefore suggests that the Shoots scheme has displayed an attractive 'figure of merit' in terms of its 'carbon footprint': comparable with that of onshore wind over their respective life-cycles.

The economics of the Shoots Barrage scheme was evaluated in some detail. This suggested that the most likely *Levelised Unit Electricity Cost* (LUEC) value was 10.8p/kWh_e (using the HM Treasury declining TDR), which is a higher figure than that obtained by DECC [11], i.e., using a discount rate of 8%. This compares with the SDC-sponsored study [12] by Black & Veatch indicated LUEC values for the Shoots tidal barrage of 3.29 p/kWh_e (with a range of 2.96-3.62 p/kWh_e) for a social discount rate of 3.5% or 6.8 p/kWh_e (with a range of 6.08-7.52 p/kWh_e) for an investor discount rate of 8%. Relative to alternative power generators, this study has confirmed the conclusions of a number of earlier studies (such as that by the consulting engineers Mott McDonald Ltd. [35]) that the electricity generated by tidal power schemes is not commercially attractive in comparison with some of the alternative technologies. Mott McDonald [35] found, using a discount rate of 10%, the 2010 LUEC for a number of power plant types: gas CCGT - 8.03 p/kWh_e; coal (without CCS) - 10.45 p/kWh_e; nuclear - 9.90 p/kWh_e; onshore wind - 9.39 p/kWh_e; and offshore wind - 16.09 p/kWh_e. However, the impact of the selected *Test Discount Rate* (TDR) was found to be significant. No allowance has been made for the cost to decommissioning the barrage in the present study; this is due to limit data being available, the longevity of the project, and also to keep this study in line with other studies, such as those by DECC [11] and the STPG [21].

ACKNOWLEDGEMENTS

This is a revised and extended version of a paper originally presented at the *8th Conference on Sustainable Development of Energy, Water, and Environmental Systems* (SDEWES) in Dubrovnik, Croatia over 22-27 September 2013 (Paper SDEWES2013.0553). This work is part of a programme of research at the University of Bath on the technology assessment of low carbon energy systems and transition pathways that is supported by a series of UK research grants and contracts awarded by various bodies. In the present context, the first author (GPH) is the Principal Investigator and joint leader a large consortium of nine university partners initially funded via the strategic partnership between *e.on UK* (the electricity generator) and the UK *Engineering and Physical Sciences Research Council* (EPSRC) to study the role of electricity within the context of '*Transition Pathways to a Low Carbon Economy*' (under Grant EP/F022832/1) over the period 2008 - 2012. The second author (CIJ) was funded in part under that grant before moving to the private sector. In 2012 the project was renewed with funding solely from the EPSRC under the title '*Realising Transition Pathways: Whole Systems Analysis for a UK More Electric Low Carbon Energy Future*' (under Grant EP/K005316/1). All the authors are grateful for the interaction with various consortium partners. Specific responses to technical queries were provided by Hayden Arrowsmith of *Lafarge Aggregates Ltd.*, Stephen Bool of *Bridgend Borough Council* in Wales (who also sat on the *Severn Barrage Steering Committee*), and Katie Gillingham of the UK *Department of Energy and Climate Change* (Office for Renewable Energy Deployment [ORED]). However, the views expressed here are those of the authors alone, and do not necessarily reflect the views of the collaborators or the policies of the funding bodies.

The authors' names are listed alphabetically.

REFERENCES

1. Alderson, H., Cranston, G. R., Hammond, G. P., Carbon and Environmental Footprinting of Low Carbon UK Electricity Futures to 2050, *Energy*, Vol. 48, No. 1, pp 96-107, 2012.

2. Parliamentary Office of Science and Technology [POST], *Electricity in the UK*, Postnote No. 280, POST, London, 2007.

3. Hammond, G. P., Energy, Environment and Sustainable Development: A UK Perspective, *Trans IChemE Part B: Process Safety and Environmental Protection*, Vol. 78, No. 4, pp 304-323, 2000.

4. Intergovernmental Panel on Climate Change [IPCC], *Climate Change 2013 - The Physical Science Basis*, Cambridge University Press, Cambridge, 2013.

5. Department of Trade and Industry [DTI], *Meeting the Energy Challenge - A White Paper on Energy*, The Stationary Office Limited, London, 2007.

6. Climate Change Act 2008, Chapter 27, The Stationary Office Limited, London, 2008.

7. Hammond, G. P., Pearson, P. J. G., Challenges of the Transition to a Low Carbon, More Electric Future: From Here to 2050 (Editorial), *Energy Policy*, Vol. 52, pp 1-9, 2013.

8. Elliott, D., Tidal Power, In Boyle, D. (ed.), *Renewable Energy: Power for a Sustainable Future*, Oxford University Press, Oxford, Third Edition, Chapter 6, pp 241-296, 2012.

9. Royal Commission on Environmental Pollution [RCEP], *Twenty-second Report: Energy - The Changing Climate*, The Stationary Office Limited, London, 2000.

10. HM Government, *The UK Renewable Energy Strategy*, Cm 7686, The Stationary Office Limited, London, July 2009.

11. Department of Energy and Climate Change [DECC], *Analysis of Options for Tidal Power Development in the Severn Estuary - Interim Options Analysis Report* (prepared by Parsons Brinckerhoff Ltd. in association with Black & Veatch Ltd.), DECC, London, December 2008.

12. The Sustainable Development Commission (SDC), *Turning the Tide: Tidal Power in the UK*, SDC, London, 2007.

13. HM Government, *Severn Tidal Power: Feasibility Study Conclusions and Summary Report*, Department of Energy and Climate Change [DECC], London, October 2010.

14. Allen, S. R., Hammond, G. P., Harajli, H. A., Jones, C. I., McManus, M. C. and Winnett, A. B., Integrated Appraisal of Micro-generators: Methods and Applications, *Proc. Instn*

Civil.   Engrs: Energy,   Vol.   161,   No.   2,   pp   73-86,   2008.

15. Hammond, G. P., Jones, C. I., Embodied Energy and Carbon in Construction Materials, *Proc. Instn Civil. Engrs: Energy*, Vol. 161, No. 2, pp 87-98, 2008.

16. Hammond, G., Jones, C., *Embodied Carbon: The Inventory of Carbon and Energy (ICE)*, Lowrie, F. and Tse, P. (eds), A BSRIA Guide - BG 10/2011, BSRIA Ltd., Bracknell, 2011.

17. El-Fadel, R. H., Hammond, G. P., Harajli, H. A., Jones, C. I., Kabakian, V. K. and Winnett, A. B., The Lebanese Electricity System in the Context of Sustainable Development, *Energy Policy*, Vol. 38, No. 2, pp 751-761, 2010.

18. Hammond, G. P., Ondo Akwe, S. S., Williams, S., Techno-economic Appraisal of Fossil-fuelled Power Generation Systems with Carbon Dioxide Capture and Storage, *Energy*,   Vol.   36,   No.   2,   pp   975-984,   2011.

19. Hammond, G. P., Harajli, H. A., Jones, C. I., Winnett, A. B., Whole Systems Appraisal of a UK Building Integrated Photovoltaic (BIPV) System: Energy, Environmental, and Economic Evaluations, *Energy Policy*, Vol. 40, pp 219-230, 2012.

20. Hammond, G. P., Howard, H. R., Jones, C. I., The Energy and Environmental Implications of UK More Electric Transition Pathways: A Whole Systems Perspective, *Energy   Policy*,   Vol.   52,   pp   103-116,   2013.

21. Severn Tidal Power Group (STPG), *Tidal Power from the Severn*, Thomas Telford, London, 1986.

22. Hammond, G. P., Winnett, A. B., Interdisciplinary Perspectives on Environmental Appraisal and Valuation Techniques. *Proc. Instn Civil. Engrs: Waste and Resource Management*,   Vol.   159,   No.   3,   pp   117-130,   2006.

23. Chapman, P. F., Energy Costs: A Review of Methods, *Energy Policy*, Vol. 2, No. 2, pp 91-103, 1974.

24. Roberts, F., The Aims, Methods and Uses of Energy Accounting, *Applied Energy*, Vol. 4, No. 1, pp 199-217, 1978.

25. Slesser, M., *Energy in the Economy*, Macmillan Press, London, 1978.

26. White, S. W., Kulcinski, G. L., Birth to Death Analysis of the Energy Payback Ratio and CO_2 Gas Emission Rates from Coal, Fission, Wind and DT-fusion Electrical Power Plants, *Fusion Energy and Design*, Vol. 48, No. 3-4, pp 473-481, 2000.

27. Woollcombe-Adams, C., Watson, M., Shaw, C. T., Severn Barrage Tidal Power Project: Implications for Carbon Emissions, *Water and Environment Journal*, Vol. 23, No. 1, pp 63-68, 2009.

28. Institut fur Energie und Umweltforschung (IFEU), *EcoTransIT: Ecological Transport Information Tool; Environmental Methodology and Data*, IFEU, Heidelburg, 2008.

29. Roberts, F., Energy Accounting of River Severn Tidal Power Schemes, *Applied Energy*, Vol. 11, No. 3, pp 197-213, 1982.

30. Heijungs, R., Guinee, J. B., Huppes, G., Lankreijer, R. M., Udo de Haes, H. A., Sleeswijk, A., *Environmental Life Cycle Assessment of Products – Guide and Background*, Report CML 92, Leiden University, Leiden, 1992.

31. Udo de Haes, H. A., Heijungs, R., Life-cycle assessment for energy analysis and management, *Applied Energy*, Vol. 84, No. 7-8, pp 817-827, 2007.

32. Brent, R. J., *Applied Cost-benefit Analysis*, Edward Elgar Publishing, Cheltenham, 1996.

33. HM Treasury, *The Green Book: Appraisal and Evaluation in Central Government*, The Stationery Office, London, 2003.

34. Moomaw, W., Burgherr, P., Heath, G., Lenzen, M., Nyboer, J., Verbruggen, A., Annex II: Methodology, In Edenhofer, O. *et al.* (eds), *IPCC Special Report on Renewable Energy Sources and Climate Change Mitigation*, Cambridge University Press, Cambridge, pp 973-1000, 2011.

35. Mott MacDonald Ltd., *UK Electricity Generation Costs: Update*, Report for the Department of Energy and Climate Change [DECC], Mott MacDonald, Brighton, 2010.

14

A Holistic ICT Solution to Improve Matching between Supply and Demand over the Water Supply Distribution Chain

Gabriel Anzaldi
Barcelona Digital Technology Center, Barcelona, Spain
e-mail: ganzaldi@bdigital.org

ABSTRACT

While many water management tools exist, these systems are not usually interconnected and therefore cannot communicate between one another, preventing Integrated Water Resources Management to be fully achieved. This paper presents the solution proposed by WatERP project[*] where a novel solution enables better matching between water supply and demand from holistic perspective. Subsystems that control the production, management and consumption of water will be interconnected through both information architecture and intelligent infrastructure. The main outcome will consist of, a web-based Open Management Platform integrating near real-time knowledge on water supplies and demand, from sources to users, across geographic and organizational scales and supported by a knowledge base where information will be structured in water management ontology to ensure interoperability and maximize usability. WatERP will thus provide a major contribution to: 1) Improve coordination among actors, 2) Foster behavioural change, 3) Reduce water and energy consumption, 4) Optimize water accountability.

KEYWORDS

SOA-MAS, Water management, Ontology, Agents, WaterML2.0, IWRM, Logical models.

INTRODUCTION

Water domain situation

In recent years, water shortage has become an increasing concern, with a growing imbalance between water demand and availability reaching critical levels. As cities grow and environmental problems escalate, managing human demand for fresh water presents an increasing challenge [1]. Increasing scarcity of supply, pollution, over-exploitation of resources and climate change are placing increasing stress on water supply systems. Meanwhile land use changes affect groundwater bodies and surface water ecosystems, putting more pressure on water reserves [2]. With ever-growing demand reaching ecological and economic limits, the need for innovative water management is acute. The worldwide gap between water demand and availability is projected to grow significantly in the next 20 years, reaching nearly 40% by 2030. In Europe, climate change is causing increased water shortages and more frequent, more severe droughts, especially in Mediterranean countries [2]. Under mid-range assumptions on temperature and

---

[*] The research leading to these results has received funding from the EC funding scheme Seventh Framework Programme FP7/2007-2013, WatERP project grant agreement n° 318603. This paper reflects only the authors' research and the Union is not liable for any use that may be made of the information contained herein.

precipitation changes, water availability is expected to decline in southern and south-eastern Europe by 10% or more in some river basins by 2030. The sectorial profile of water abstraction is also expected to change [3]. Meanwhile, water demand is increasing as a result of population growth, changing consumer patterns and growing industrial use [4]. In order to secure water supplies into the future, there is an urgent need to transition towards a more water-smart society and develop water-wise solutions to improve water and energy efficiency, reduce water consumption and preserve water resources [2].

Water resource management situation

Water resource management involves a wide array of actors, from water authorities to water regulators, water utilities and finally the end-users. While many optimization, planning and monitoring tools have been developed and are currently used, such as hydro meteorological forecasting and hydrologic and hydraulic models, decision support systems for reservoir and hydraulic infrastructure operations, and real-time monitoring and control systems for water treatment and distribution, these systems cannot communicate between one another and currently no framework is available for integrating all of these applications [5]. Yet water management is becoming increasingly more complex, with continual changes in human and natural systems affecting water availability, access, affordability and quality [4] and therefore, there is a need for more integrated and adaptive management approaches based on reliable monitoring systems and a solid knowledge base [2].

Although water and energy savings have been achieved in various sectors, these improvements are localized and uncoordinated. Each entity currently acts independently without much knowledge regarding the needs, constraints or operations of the others and information is not easily accessible. Yet net water savings and environmental improvements can only be realized if the water saved in one area is not used elsewhere by others or downstream [6]. In order to achieve wide scale improvements, there is a need for enhanced coordination, cooperation between water supply actors across different scales, in order to address both long-term water imbalances (water scarcity), and enhance resilience to drought [7].

In parallel, there is a need for increased information sharing. If information were shared among the various decision-makers and stakeholders, operations could be coordinated, better decisions could be made, water supplies could be prioritized according to needs and changing conditions, overall water use efficiency could be improved, and water shortages and energy waste could be reduced.

Why this is the right moment to go a step forward?

Over the past decades, Europe has made important progress in regards to infrastructure, technologies and water management. However, despite substantial efforts and improvements related to water resources management, the 2010 European Environment State and Outlook Report revealed that many of the water bodies will fail to meet the Water Framework Directive (WFD) objectives of achieving good status by 2015. Meanwhile, freshwater systems are still under pressure, demand often exceeds availability, and drought and water stress are expected to increase as a result of climate change. There is a need for a more sustainable approach for water resources management to improve water demand management more widely across Europe and to avoid mismanagement of water resources, especially in areas of water scarcity [8].

While the WFD river basin management plans will remain the primary framework for managing water resources in Europe, a new demand-side management approach is needed, including measures such as water pricing and efficiency in order to secure water for all essential uses. This new water paradigm is what the upcoming Blueprint is set out to accomplish. In parallel, the 2011 Flagship initiative on resource efficiency under the EU 2020 Strategy makes water saving measures and increasing water efficiency a priority.

One of the most innovative elements of the WFD is its integrative approach to water management, bringing together different water management issues within a unified framework. Recent ICT technological advances make sharing information and such integration all the more possible. Information systems are now being called upon to support knowledge management and not just to process data or information. Today, information can be exchanged in real-time, very large amounts of data can be processed in automated manners and web-based internet services enable information to be collected, processed and shared in ways that were not possible before. With the upcoming next generation of semantic applications, data can become more machine-interpretable by developing ontologies that can support the development of integrated software systems, and by aligning the ontologies of different applications, information can be shared, resulting in increased interoperability [9]. A framework providing interoperability between loosely coupled software applications and data sources, as is being proposed in WatERP project, will enable integrated water resources management to be achieved. Indeed with new applications and web-based services supplementing existing tools within a joint collective network, current water resource management capabilities can be greatly enhanced.

One of the first ICT water management resource planning concepts was adopted by Visseman ad Welty [10] in 1985 and the next year, the first water resource planning tool was implemented and integrated into a Decision Support System called StateMod [11], it is capable of making a comparative analysis for the assessment of various historic and future water management policies, simulating water flows, allowing reservoirs to be operated with multiple accounts serving multiple users, and estimating natural stream flows and reservoir data. During following years, the interest in the "systems" approach grew with the advent of much more user friendly PC-based software. The evolution of PC optimization and mathematical tools permitted the application of these technologies in the water resource management planning framework. However, the process of translating a water resources problem into a mathematical problem causes much of the reality of the problem to be lost. As a result, analysts turned their attention to more user-focused descriptive approaches. Expert systems and emerging artificial intelligence techniques offered the promise of a more user-centred approach. User-defined heuristic decision rules [12] were developed, the resulting advantage of this architecture being the inclusion of the clients in the problem definition to solve and adapt the water resource planning management according to their necessity. Early on, the experience with experts systems indicated that the most important need for decision support systems was a user-friendly database management system.

In the last decade, ICT management architectures for complex installations (water deposit, water treatment plants, waste water treatment plants, water distribution, etc.) have become widespread, focusing on concrete and specific applications. These infrastructures provide information about the status and performance of the installation, integration of sensors, media and databases [13-17]. Numerous efforts in artificial intelligence were made to solve problems of conventional processes by applying different knowledge-based systems. Related to the water supply distribution chain, knowledge techniques have been developed as isolated support systems for monitoring, diagnosis, design, process optimization, etc., [14, 18-20] among others. These systems are becoming more and more sophisticated, advanced and have enabled significant water management improvements. However, in today's complex installations, the information is treated as a local resource and is not shared between systems.

THE SOLUTION

Holistic view, basin scale approach

Water resources management is extremely complex, related not only to the environment but also to the numerous human activities that are carried out within this environment. Water availability and usage depend on the timing and manner of its arrival (rainfall intensity, rain or snow, duration, frequency), the physical setting of the region (climate and weather, topography, geology), the engineering structures in place, the environmental constraints (existing ecosystems), as well as the legal regulatory context and institutional policies.

Traditional approaches to water resources management have typically been based on black-box optimization models, handled by technical people and developed for very specific purposes. They do not include interactions with the end users or stakeholders involved and have not been able to include in their computations the full variety of important factors that must be considered by decision makers, or in ways that are transparent to the public.

The water industry lacks an adequate holistic understanding of water supply, its use, and how it flows. A common subject across the entire water chain is the need for improved data collection and the transformation of that data to generate actionable information and knowledge. Moreover, the Water industry is under pressure to take a more holistic perspective of water, considering its whole life cycle from abstraction to treatment, distribution, use and end treatment. This also means a stronger recognition of the role of green infrastructure.

The potential for service integration between water sectors is enabling companies operating in the water industry to examine mechanisms by which technology and its usage can bring holistic improvements to the water network and bring potential reductions to their operating costs.

Therefore, there is an urgent need for more holistic approaches that can address the complex coupled human and physical system interactions at the basin scale. Furthermore, new integrative approaches must include multi-resolution capacity so that findings and information can be transferred and used across models and users, based on an agreed-upon conceptual model of the system. Doing so will help stakeholders and decision-makers understand what are the main issues and challenges at the system level, but also for each stakeholder.

Scope

The WatERP solution focuses on the different actors involved in water supply distribution chain and on obtaining, from each one of these, the necessary parameters required for enabling demand to be matched with supply across the entire cycle. For this purpose, WatERP will provide standard interfaces to integrate the necessary information from each supply management step, either through direct interaction with control or management systems.

Ultimately, WatERP will result in a web-based Open Management Platform (OMP) tool that integrates near real-time knowledge on available water supplies and demand, from water sources to users, and across geographic and organizational scales, so the information from each step of the process can be exchanged and accessed and the entire water supply distribution network can be viewed, understood and improved in an integrated and collaborative manner. This platform will be integrated in an intelligent ICT architecture that interconnects different management tools (or building blocks) available in the supply distribution network and will be supported by an ontology driven knowledge base on water supplies and water usage providing a continuous flow of information including historical, current, and forecasted values.

Description

The OMP will provide water resource managers inferred information regarding water supplies, flows, water consumption patterns, water losses, distribution efficiency, and water supply and demand forecasts, within an intelligent unified framework based on open standards. This information will be stored using semantics and common language which will be defined in the water management ontology to ensure interoperability and maximize usability. External linkages to costs, energy factors, control systems, data acquisition systems, external models, forecasting systems and new data sources will be made possible for easy integration into the system. The main purpose of this information interaction and processing will be to improve the matching between supply and demand. To achieve this final goal, tools will be developed to support coordination of actions throughout the entire water supply distribution chain, prioritization of water uses, distribution efficiency improvements, and water, energy and cost savings. In addition to the openly extensible technology platform, WatERP will provide end-to-end consulting and systems integration services that include:

- Operational dashboards for continuous monitoring of time-sensitive key indicators and metrics;
- Advanced rules management, constraint-based optimization, and visualization tools to more effectively manage and automate the water management decision making processes;
- Integrated high-resolution local weather predictions that will enable optimizing weather sensitive water management operations to improve availability;
- Innovative capabilities for standards-based, secure data exchange;
- Analytical demand-management and decision-support tools as well as access to information from other sources;
- Ensuring appropriate information is available at the right time, place and scale.

To accomplish its function, WatERP will develop the following building blocks:

- Decision Support System (DSS);
- Demand Management System (DMS);
- Water Data Warehouse (WDW).

The information produced for these building blocks and others systems (external systems) will be interconnected through a specific ICT architecture, the knowledge structured and managed by the water management ontology and the interaction of the information and knowledge with the water resource manager performed in the OMP. The following sections of this paper are focused on the last three systems.

The ICT architecture

The communication architecture will focus on providing intelligent and near real-time linkage between the various water supply distribution chain management tools or building blocks (Data Management Systems, Decision Support systems, Demand Management Systems, Weather Forecast Systems, etc.) and the OMP that will support water management decisions, enabling knowledge-based water governance to be achieved throughout the water supply distribution chain.

A combined Service Oriented Architecture (SOA) with Multi-Agent System (MAS) is being designed to:

- Link each decisional/informational system to help the integration in a collaborative framework;
- Provide near real-time information flow;
- Distributed intelligence to generate actions and alerts related to management processes;
- Procedure to perform orchestration of existing and new management tools throughout the whole architecture.

Nowadays, SOA architecture is a booming technology with a high level of maturity and success. It is widely used to exchange information between systems that are located remotely and managed by third-parties (e.g., legacy systems, systems with unknown codification and exploitations, etc.). This architecture permits the orchestration and automation of critical processes using a distributed architecture that exchanges the information in a standardised way, XML being the format most typically used. Thanks to this concept, the internet provides millions of services and resources around the world. A good example in the water domain was developed by CUAHSI (http://www.cuahsi.org/), providing hydrological information via Web Services.

Nowadays, the problem is not the lack of information, but rather its integration with a common goal. With the aim of integrating and reusing knowledge provided by different services and resources, techniques such as BPEL engines [21] or matchmaking multi-agent systems [22] are used. The BPEL engine allows creating static business processes with the services to orchestrate those [22-24]. Alternatively, the MAS are used for their flexibility and dynamicity in matchmaking problems. The MAS can solve conflicts, adapt to changes and is highly scalable [25].

Water supply management involves a very large quantity of control and management systems (services) that must be interconnected and orchestrated, along with the emergence of new services. WatERP's SOA-MAS architecture is based on a pool of services provided by the building blocks and ontological instantiation. These services should be orchestrated with the purpose of integrating all information and facilitating the decision making to improve water resource management and energy efficiency. Different alternatives exist for this purpose such as previously-mentioned BPEL and MAS. In spite of its extended application, BPEL presents a well-known disadvantage, its stiffness [26]. This disadvantage limits its application in the WatERP project, mainly because of its lack of flexibility and inconvenience for easy integration of new services or the modification of existing ones.

To overcome this challenge, one of the most applied solutions by the scientific community is, the implementation of a matchmaking process by using agents which auto-manages services in order to fit the needs [22-24]. The MAS orchestrator is flexible to integrate new services at any time which is essential in the water management. This flexibility is provided by the auto-organization of the agents which manages the needs of the platform and the knowledge of the provided services. Moreover, the use of MAS provides numerable intrinsic benefits such as fault tolerance, scalability and flexibility. Therefore, MAS is the piece of the puzzle which best fits the different services because it provides two important benefits to the orchestration which are not provided by the BPEL. These benefits are: flexibility and scalability.

In order to ensure interoperability all the ICT architecture is designed following standardization principles. The WatERP MAS is therefore being designed by following standardized languages to communicate among agents such as: FIPA-ACL and KQML.

Once the MAS orchestration identifies the service provided by a building block that satisfies the requirement, the involved agent need to interact with the deployed services to transfer the information/knowledge required. To do so, agents and building blocks must accomplish an open interface based on a standard layer to provide its functionalities. This standard layer facilitates the discovering of service functionalities and is based on Open Geospatial Consortium (OGC®, http://www.opengeospatial.org/) WPS/WFS specifications. Moreover, with the aim to standardize the exchange information between the MAS and services, OGC®'s WaterML 2.0 will be used as the common transport format.

Figure 1 depicts the ICT architecture design. The three main building blocks represented (DMS, DSS and External Systems) interchange information with the MAS through an open interface based on WPS (or WFS) standard schema. The integration of MAS with WPS/WFS conforms the SOA-MAS architecture that enables interoperability by connecting each of the building blocks according its requirements and needs (matchmaking process). Furthermore, the figure shows how the information from water operators (Authorities/Utilities) is gathered, transformed and published on an OGC-SOS server towards feeding the Water Data Warehouse (WDW). The WDW offer the data/information to the visualization and decisional systems such as DMS, DSS, External systems and the OMP by using the same architecture. Transversally to this architecture where the information flows continuously, the WatERP ontology permit to enhance semantically the water domain knowledge by adding metadata information related with water domain decisional, observation and measurement process. This semantically definition stored in the ontology is able to improve the interoperability by enhancing data provenance (by categorising it measurement process) and data fusing (by understanding the measurement nature).

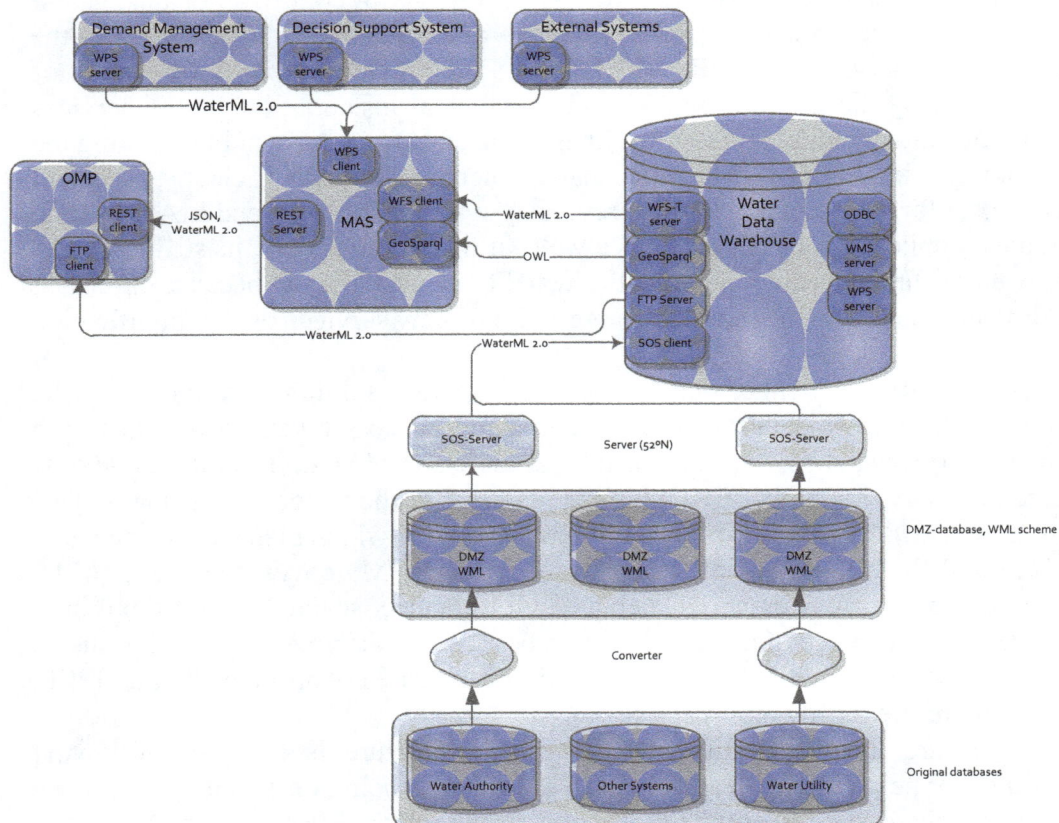

Figure 1. ICT architecture initial design

The water management ontology

The WatERP ontology (Figure 2) will provide easy access to information and will support decision-making and water resource planning. All of the decision making processes will be vertically integrated using universal standards in such a way that different multilevel inferences can be made:

- In each step of the water supply distribution chain;
- From the interactions among functions involved in each step of the water transport from the hydrometeorological data to the final user;
- From the interplay between currently separated control and optimization systems such as reservoir or hydroelectric plant decision support systems or water treatment and distribution management tools;
- From analysis of the impact of water savings on energy savings.

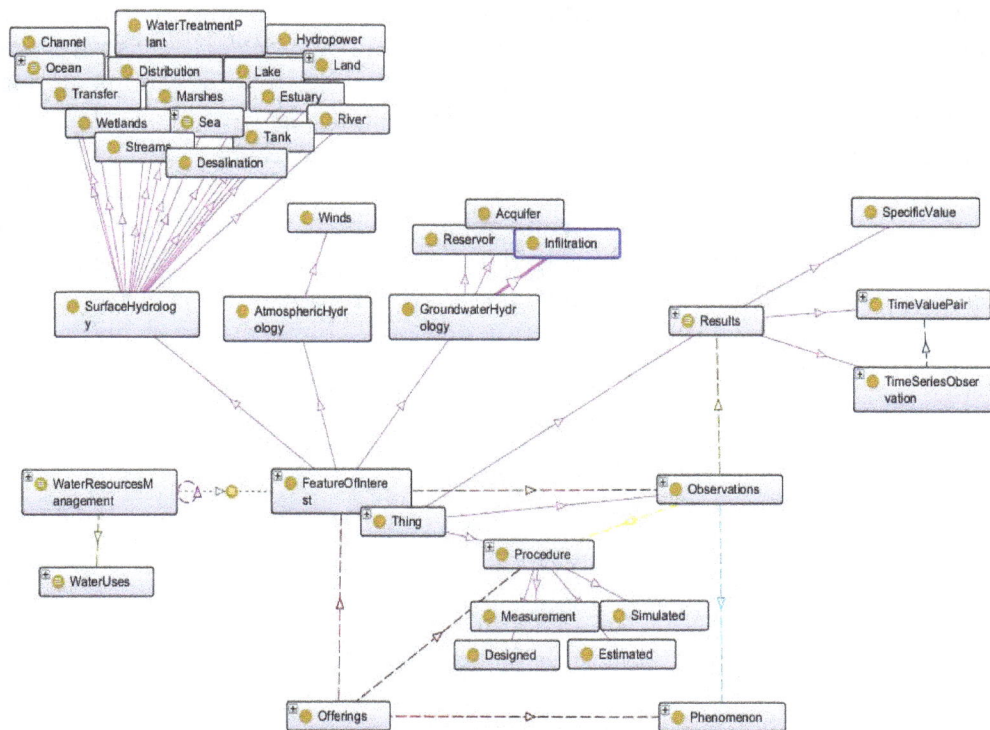

Figure 2. WatERP ontology general overview

The water management ontology contains corresponding ontologies, as well as multi-level data regarding:

- Water resource availability;
- Ecological, cultural and social functions of water resources and potential impacts of changes on hydrological regimes;
- Current water infrastructure/assets and the economic value of water;
- Administrative, policy or regulatory issues of relevance;
- Sectorial use and water hierarchy.

Existing ontologies such as those developed by CUAHSI and the OGC® have tried to model the hydrologic cycle from the hydrological and environmental perspective. Following a different approach, the National Aeronautics and Space Administration (NASA) tried to give a definition of the hydrological cycle with the perspective of the earth/environment science connection (NASA-SWEET ontology). On the one hand, the knowledge base such as the one developed by CUAHSI was done by merging the hydrological and environmental fields in a way as to provide sensor information of the water environment and correlate it with environmental data. However, CUAHSI's application does not show linked information semantically or link it with the web to create a certain meaning for each data. On the other hand, ontologies such as SWEET, OGC® (Onto Sensor and more) and W3C-Semantic Incubator Group (SSN ontology) provide mechanisms that support information retrieval from sensors (case of Onto Sensor and SSN ontology) and also provide an environment where multiple fields and sciences are linked semantically and are defined by a trustworthy organization (case of SWEET).

However, there is no ontology that encompasses the water cycle from its management perspective with the aim to establish recommendations and alerts regarding actions taken or decisions made concerning its various elements. Such harmonization could provide the water field the possibility of an enhanced understanding, and in an automatic way, of water resource management systems. The novelty of the WatERP ontology over previous developments lies in its inclusion of man-made infrastructure elements. Doing so enables them to be linked to the natural water flow processes such that the interactions among natural and human made entities can be better understood, along with their effects on water resources management. These human made modifications of the natural hydrological cycle flow paths have been defined in WatERP as human-altered paths. As a result, the WatERP ontology can semantically represent both the human-altered and natural paths to discover new interrelations (hidden knowledge) among water resource elements, ultimately enabling improvements and new strategies for water resource management (Figure 3).

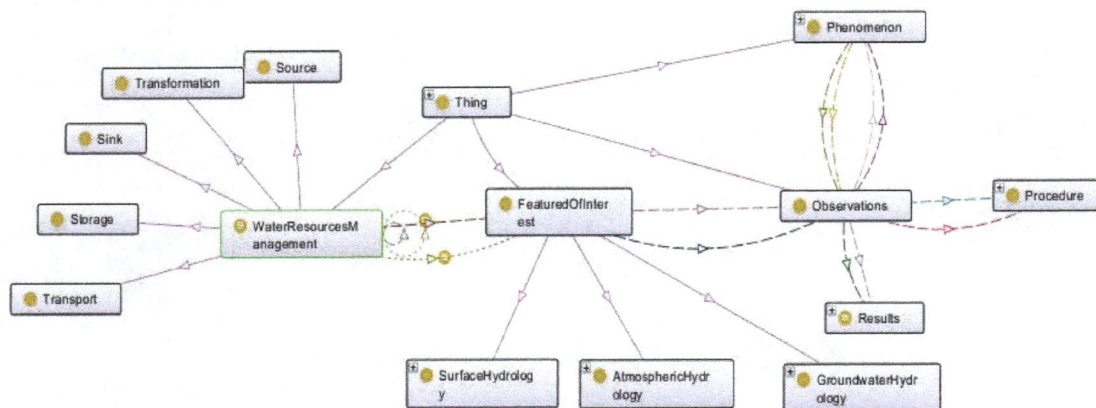

Figure 3. WatERP ontology decisional process representation (logical models) and the linkage with the observation and measurement process

This knowledge representation is supported by a data provenance mechanism in order to define a process towards observation understanding and standardization of the ontology including concepts and standard terms provided by other ones, such as the NASA, CUAHSI, OGC® and World Wide Web Consortium (W3C) ontologies. Moreover, the WatERP ontology has been constructed following the principles of Linked Open Data Cloud (LODC) contributing to the aim of achieving interconnectivity within the WatERP solution. LODC permits resources to be accessed by an URI and linked with other elements enabling automatic understanding (human-readable). From this, ontological information can be integrated and accessed, the terms of different vocabularies can be mapped, and data fusion supported. All of these features contribute to resolve data conflicts by integrating data from different sources into an entity.

The Open Management Platform (OMP)

The WatERP Open Management Platform will be a water supply distribution chain information hub, which will support decision making at different stages such as water supply and demand (and forecasting), determining water consumption patterns, water allocation decisions, etc., all of which will help improve overall water governance. The OMP will integrate the outcomes of the existing building blocks and modules in a graphical way which will empower local and global management. Along this line of improvement, a business intelligence tool will also be provided including predefined dashboards and reports according to the needs. These dashboards and reports will be defined in a way that they can be easily extended or modified according to specific needs.

It is important to remark that WatERP innovates with respect to the existing hydrological systems knowledge representations by defining an ontology that encompasses all of the elements and knowledge involved in water management. The Water Management Ontology is specified using logical models, which constitutes the basis for the analyses performed by the water resource manager. A logical model is the representation of the managed water supply system components and relations that acts as interface between the water manager and the water management ontology. As such, the user can obtain information from specific ontology as well as recommendations based on applying a rule-based analysis based on expert experience. This will allow two kinds of abstractions to be made: one regarding to the elements' interactions and other regarding to the specific information available in each of the physical elements involved. In the OMP, the abstraction of the geographical environment is achieved by the inclusion of the logical model.

In practice, the inclusion of logical models allows parts of the water supply distribution system to be grouped or detailed as needed in function of the quantity and quality of existing information and the decisions that must be made. Thus, it is possible to include in the platform all decision groups (one or various elements grouped) of the system with the same level of knowledge than the users, so that the system can grow over time as the user gains more knowledge about the physical environment. The introduction of logical models does not mean that the system removes any reference to geographic elements or temporal scope of decision, but rather, that each of them are defined according to the type of logic element to which they relate:

- Each element of the logical model corresponds to one or a set of physical elements;
- Each element of the logical model has its corresponding temporal scope. It should be noted that there may be other logical elements containing the same information but with a different temporal scale.

Logical models (Figure 4) define each use case which is managed by the resource manager. Therefore, it is essential that the OMP implements a Graphical User Interface (GUI) for handling, creation and editing of the logical models. In addition to allowing visualization of the water management chain where decisions are being made, the GUI has to permit access to the information contained in each entity of the model.

For interoperability of logical models in the field of hydrology, the only significant existing effort is HY_Features, However, at present there is not yet an OGC® standard (under Discussion Paper state) for this and its technology is not enough mature to use it at this moment. Because of this, logical models are actually instantiated entities of the WatERP management ontology; and the proposed encoding and exchange language of logical models is the same used for the actual design of the ontology (typed in an OWL format). It is important to mention that OGC® is currently working on the development of a future standard for semantic representation in the water domain (HY_Features). This initiative is in its early stages of development; nevertheless, it is being considered and will be tracked during the design of the WatERP management ontology.

Any logical model has a correspondence to a physical model. A physical model is a collection of real elements that match a structure consisting of a geographical positioning component and other associated information (geodata). These elements define how things work in real life. Therefore, they have to be also managed by the OMP facilitating interaction with their information. As the standard language for interoperability with geodata, the use of Geography Markup Language (GML) is used, both standard Open Geospatial Consortium and ISO 19136:2007.

Logical and physical models contain information to be provided through OMP to the water resource manager, such as observational data that will be used as input to decision-making processes. This information is based on time series, and these series are within the scope of hydrology. There have been several efforts to standardize the exchange of this kind of information, these efforts have concluded in the specification of WaterML2.0, which is an OGC® encoding standard for the representation of hydrological observations data with a specific focus on time series structures. WaterML2.0 is implemented as an application schema of the Geography Markup Language version 3.2.1 (GML), making use of the OGC® Observations and Measurements standards. WaterML2.0 transports can be made in any form: email, ftp, file-copy, arbitrary http or standardized http transfer (as OGC® Sensor Observation Service or Web Feature Service), this constitutes a very important characteristic for information exchange.

In summary, the GUI concerning the decision making process exploits ontological resources defined in the specific ontology (using logical models) to give the user information linked semantically and associated with its geographical location and temporal scale. This linkage between concepts (ontological resources), temporal scale and geographical information renders graphically understandable measured variables and facilitates decision making and permits to implement practically the holistic approach.

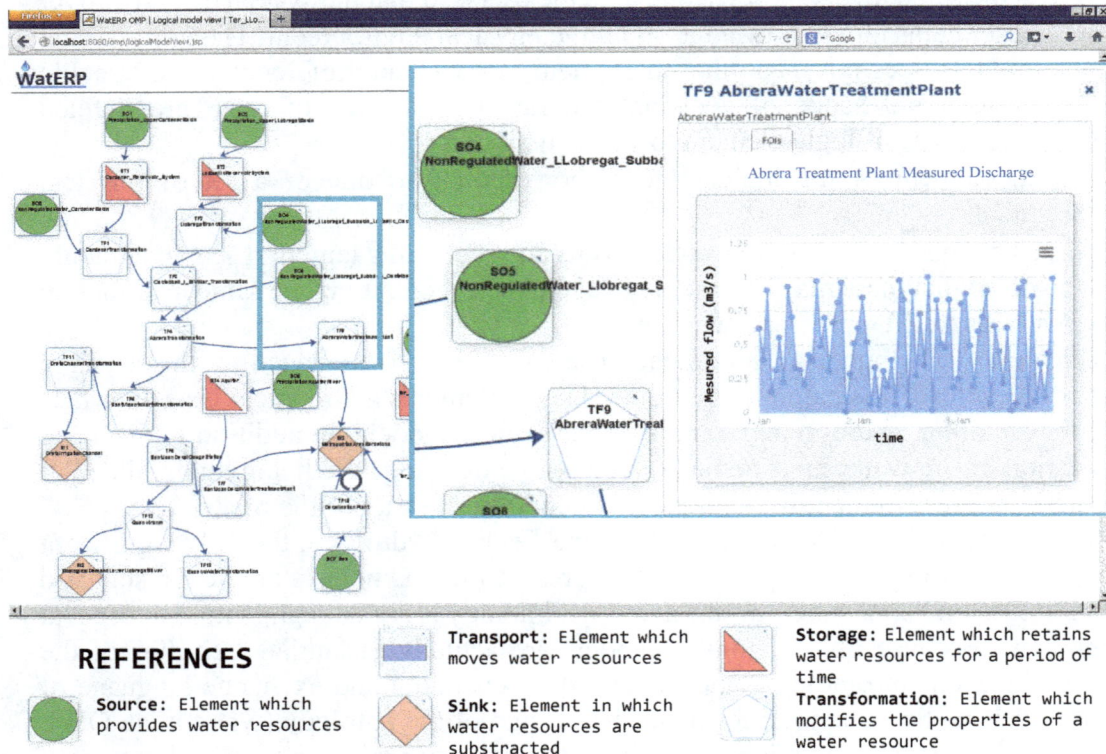

Figure 4. Logical model representation inside the OMP

CONCLUSIONS

The combination of SOA and MAS architectures and particularly the matchmaking use of agents, provides the whole architecture with:

- The capacity of selecting the most appropriate and required information for the water resource manager (via OMP) and the rest of modules interconnected;
- Conflict resolution between information providers;
- Orchestration of the architecture modules in order to manage the knowledge provisioning according to the desired business process goals (e.g., energy efficiency, matching supply with demand, etc.);
- Common and standardized language that permits the communication between agents (e.g., using FIPA language; KQML) and with the building blocks of WatERP's architecture that have a service behavior (e.g., WaterML2.0). Given the interoperable nature of the proposed system and the use of open standards, additional data sources, such as future smart metering expansions, can be easily discovered and added as they become available.

The ontology driven knowledge-management has been designed with the aim to support domain expert users in their decision making process and enhance the comprehension of the water supply systems upon which it is applied. The Water Management Ontology has been based on defining the necessary ontological resources that describe water managers' expertise related to managing water supply and demand. As previously pointed out, the WatERP project extends existing hydrological ontologies by including man-made infrastructure elements and enabling semantic representations of the human-altered water flow paths and therefore permitting new discoveries to be made for water resources management.

Until now, water supply and distribution management has been limited to isolated, uncoordinated solutions. The WatERP Open Management Platform will empower these existing solutions while at the same time offer a new water supply distribution management tool. This management tool, holistic in its approach, will enable global benefits (in terms of water optimization and reducing water and energy consumption) instead of only local ones. This is very much in line with the WFD objective of attaining integrated water resources management (IWRM). As the information will be treated from a higher point of view, it will facilitate the coordination between the different stakeholders, making them participants in the global situation, and recommending correct actions to be taken.

At the river basin level, the OMP will help improve water supply allocations among users and needs, by making information regarding available water supplies and demand more accessible. From this, water usage hierarchies could be established, available water supply sources could be prioritized, river basin scale water balances could be made, sectorial water usage could be better quantified, and illegal abstractions could be identified. In addition, the platform will enable greater cooperation among water regulators, operators and users which will lead to significant water savings.

At the distribution-network level, the information provided by the WatERP Open Management Platform will enable real-time tracking of water supplies, flows and distribution efficiency across the entire distribution network. This will enable daily operations to be better coordinated, water energy savings to be identified, ultimately resulting in water savings and increased overall efficiency. Regarding short-term benefits, a better demand prediction as well as demand and network monitoring in (near) real-time, combined with more accurate supply data and forecasts, will allow for a more energy-efficient distribution-network operation (helping determine which reservoirs or tanks should be filled and when, which pumps should be started and when, etc.).

Finally, the proposed WatERP Open Management Platform encourages open data policies and supports current standardization efforts to develop both European and international standards for water data sharing by fuelling further developments and by providing valuable feedback from the testing of existing standards in real-life scenarios, including how to address data security and data quality issues as well as how to properly track the treatment processes applied to the data, in particular, hydro-meteorological data.

REFERENCES

1. Brandes, O. M., Ferguson, K., M'Gonigle, M., Sandborn, C., At a Watershed: Ecological Governance and Sustainable Water Management in Canada, *Victoria: POLIS Project on Ecological Governance*, University of Victoria, 2005.
2. EC (2011): Commission Recommendation on the Research Joint Programming Initiative 'Water Challenges for a Changing World', 27 October 2011, *COM (2011) 7403 final*, Brussels, 2011.
3. EC (2007): Communication from the Commission to the European Parliament and the Council on addressing the Challenge of Water Scarcity and Droughts in the European Union, *COM (2007) 414 final*, Brussels, 2011.
4. Aquawareness Policy Forum 2010 Final Report, Water 2030 - Who Cares?, *European Water Partnership*, Brussels, 2010.
5. Schevers, H., Drogemuller R., Semantic Web for an Integrated Urban Software System, *MODSIM Conference*, Melbourne, Australia, 2005.
6. Dworak, T., et al., EU Water Saving Potential, ENVD2/ETU/2007/0001r, *Ecologic-Institute for international and European Policy*, 2007.
7. Kampragou, E., Apostolaki, S., Manoli, E., Froebrich, J., Assimacopoulos, D., Towards the Harmonization of Water-related Policies for managing Drought Risks across the EU, *Environmental Science Policy*, Vol. 14, pp 815-824, 2011.

8. EC (2008): Follow up Report to the Communication on Water Scarcity and Droughts in the European Union, *SEC(2008) 3069*, Brussels, 2008.
9. Schevers, H., Trinidad, G., Drogemuller, R., Towards Integrated Assessments for Urban Development, *Journal of Information Technology in Construction*, 2006.
10. Visseman, W., Welty, C., *Water Management: Technology and Institutions*, New York: Harper and Row, 1985.
11. Parsons, R., Bennett, R., Reservoir Operations Management Using a Water Resources Model, *Proceedings of Operations Management* 2006 conference, pp 304-311, 2006.
12. Koch, H., Grünewald, U., A Comparison of Modelling Systems for the Development and Revision of Water Resources Management Plans, *Water Resources Management* Vol. 23, pp 1403-1422, 2009.
13. Beran, B., Piasecki, M., Engineering new paths to water data, *Computers & Geosciences* Vol. 35, pp 753-760, 2009.
14. Cetinkaya, C. P., Fistikoglu, O., Fedra, K., Harmancioglu, N. B., Optimization Methods applied for Sustainable Management of Water-scarce Basins, *Journal of Hydroinformatics*, Vol. 10, pp 69-95, 2008.
15. Chau, K. W., An Ontology-based Knowledge Management System for Flow and Water Quality Modelling, *Advances in Engineering Software*, Vol. 38, pp 172-181, 2007.

16. Cortés, U., Sànchez-Marrè, M., Sangüesa, R., Comas, J., R.-Roda, I., Poch, M., Riaño, D., Knowledge Management in Environmental Decision Support Systems, *AI Communications*, Vol. 14, pp 3-12, 2001.

17. Cortés, U., Martinez, M., Comas, J., Sànchez-Marrè, M., Rodriguez-Roda, I., A Conceptual Model to facilitate Knowledge Sharing for Bulking Solving in Wastewater Treatment Plants, *AI communications*, Vol. 16, pp. 279-289, 2003.

18. Liu, S., Supply Chain Management for the Process Industry, Doctoral Thesis at University College, London, 2011.

19. Westphal, K. S., Vogel, R. M., Kirshen, P., Chapra, S. C., Decision Support System for Adaptive Water Supply Management, *Water Resources. Planning. Manage.*, pp 129-165, 2003.

20. Aulinas, M., Nieves, J. C., Cortés, U., Poch, M., Supporting Secision making in Urban Wastewater Systems using a Knowledge-based Approach, *Environmental Modelling & Software*, Elsevier, 2011.

21. Puttonen, J., An Application of BPEL for Service Orchestration in an Industrial Environment, *Emerging Technologies and Factory Automation*, pp 530-537, Hamburg, 2008.

22. Foner, L., A Multi-Agent Referral System for Matchmaking, *PAAM96 Proceedings*, London, England, 1996.

23. Ebrahimi, B., Bertels, K., Vassiliadis S., Sigdel K., Matchmaking within Multi-agent Systems, *Proceedings of ProRisk 2004*, Vendhoven, pp 118-124, Nethederlands, 2004.

24. Maximilien, E., Singh, H., Multiagent System for Dynamic Web Services selection, *Proceeding of first Workshop on Service-Oriented Computing and Agent-Based Engineering (SOCABE at AAMAS)*, pp 25-29, 2005.

25. Park, A. H., Park, S. H., Youn, H. Y., A Flexible and Scalable Agent Platform for Multi-Agent Systems, *International Journal of Engineering and Applied Sciences 4:1*, 2008.

26. Mehdi, S., Ralyté, J., Jean-Henry, M., Business Process Flexibility in Service Composition, *Exploring Service Science*, pp 158-173, Geneva: Springer, 2011.

Permissions

All chapters in this book were first published in JSDEWES, by SDEWES Centre; hereby published with permission under the Creative Commons Attribution License or equivalent. Every chapter published in this book has been scrutinized by our experts. Their significance has been extensively debated. The topics covered herein carry significant findings which will fuel the growth of the discipline. They may even be implemented as practical applications or may be referred to as a beginning point for another development.

The contributors of this book come from diverse backgrounds, making this book a truly international effort. This book will bring forth new frontiers with its revolutionizing research information and detailed analysis of the nascent developments around the world.

We would like to thank all the contributing authors for lending their expertise to make the book truly unique. They have played a crucial role in the development of this book. Without their invaluable contributions this book wouldn't have been possible. They have made vital efforts to compile up to date information on the varied aspects of this subject to make this book a valuable addition to the collection of many professionals and students.

This book was conceptualized with the vision of imparting up-to-date information and advanced data in this field. To ensure the same, a matchless editorial board was set up. Every individual on the board went through rigorous rounds of assessment to prove their worth. After which they invested a large part of their time researching and compiling the most relevant data for our readers.

The editorial board has been involved in producing this book since its inception. They have spent rigorous hours researching and exploring the diverse topics which have resulted in the successful publishing of this book. They have passed on their knowledge of decades through this book. To expedite this challenging task, the publisher supported the team at every step. A small team of assistant editors was also appointed to further simplify the editing procedure and attain best results for the readers.

Apart from the editorial board, the designing team has also invested a significant amount of their time in understanding the subject and creating the most relevant covers. They scrutinized every image to scout for the most suitable representation of the subject and create an appropriate cover for the book.

The publishing team has been an ardent support to the editorial, designing and production team. Their endless efforts to recruit the best for this project, has resulted in the accomplishment of this book. They are a veteran in the field of academics and their pool of knowledge is as vast as their experience in printing. Their expertise and guidance has proved useful at every step. Their uncompromising quality standards have made this book an exceptional effort. Their encouragement from time to time has been an inspiration for everyone.

The publisher and the editorial board hope that this book will prove to be a valuable piece of knowledge for researchers, students, practitioners and scholars across the globe.

List of Contributors

Martin A. Hendel
Paris City Hall, Water and Sanitation Department, F-75014, Paris, France
Université Paris-Est, Lab'Urba, EA 3482, EIVP, F-75019, Paris, France
Univ Paris Diderot, Sorbonne Paris Cité, MSC, UMR 7057, CNRS, F-75013, Paris, France

Morgane Colombert
Université Paris-Est, Lab'Urba, EA 3482, EIVP, F-75019, Paris, France

Youssef Diab
Université Paris-Est, Lab'Urba, EA 3482, EIVP, F-75019, Paris, France

Laurent Royon
Univ Paris Diderot, Sorbonne Paris Cité, MSC, UMR 7057, CNRS, F-75013, Paris, France

Anni Koci Kallfa
Department of Biotechnology, Faculty of Natural Sciences, University of Tirana, Albania

Fatbardha Babani
Department of Biotechnology, Faculty of Natural Sciences, University of Tirana, Albania

Ariana Ylli Kraja
Department of Biotechnology, Faculty of Natural Sciences, University of Tirana, Albania

Thomas J. Voltz
Division of Water Sciences, University of Applied Sciences (HTW) Dresden, Germany

Thomas Grischek
Division of Water Sciences, University of Applied Sciences (HTW) Dresden, Germany

Marcel Spitzner
Faculty of Electrical Engineering, University of Applied Sciences (HTW) Dresden, Germany

Jana Kemnitz
Faculty of Electrical Engineering, University of Applied Sciences (HTW) Dresden, Germany

Rudolf Irmscher
Executive Management, Stadtwerke Heidelberg GmbH, Heidelberg, Germany

Ramchandra Bhandari
Institute for Technology and Resources Management in the Tropics and Subtropics, Cologne University of Applied Sciences, Betzdorfer Strasse 2, 50679 Cologne, Germany

Anita Richter
Deutsche Gesellschaft für Internationale Zusammenarbeit (GIZ) GmbH, Potsdamer Platz 10, 10785 Berlin, Germany

Andre Möller
Alteburger Strasse 109, 50678 Cologne, Germany

Rolf.-P. Oswianoski
Virtu Consult UG, Handjerystraße 94, 12159 Berlin, Germany

Ying Miao
Centre of Development Studies, University of Cambridge, Cambridge, United Kingdom

Yu Jia
Department of Engineering, University of Cambridge, Cambridge, United Kingdom

Dušan Polomčić
Department of Hydrogeology, Faculty of Mining and Geology, University of Belgrade, Đušina 7, Belgrade, Serbia

Dragoljub Bajić
Department of Hydrogeology, Faculty of Mining and Geology, University of Belgrade, Đušina 7, Belgrade, Serbia

Jelena Zarić
Department of Hydrogeology, Faculty of Mining and Geology, University of Belgrade, Đušina 7, Belgrade, Serbia

Muriel de Oliveira Gavira
School of Applied Sciences, University of Campinas, Brazil
Department Faculty of Sciences, University of Lisbon, Portugal

Amy Tang
Civil Engineering and Engineering Mechanics Department, Columbia University, New York, USA

John E. Taylor
Charles E. Via, Jr. Department of Civil and Environmental Engineering, Virginia Tech, Blacksburg, USA

Thomas Weiss
Electrical Power Systems, Helmut Schmidt University, Hamburg, Germany

Karl Zach
Energy Economics Group, Technical University of Vienna, Austria

Detlef Schulz
Electrical Power Systems, Helmut Schmidt University, Hamburg, Germany

Larisa Jovanović
ALFA University, Belgrade, Serbia

Lazar Cvijić
ALFA University, Belgrade, Serbia

Karl-Heinz Kettl
Institute for Process and Particle Engineering Graz University of Technology, Graz, Austria

Nora Niemetz
Institute for Process and Particle Engineering Graz University of Technology, Graz, Austria

Michael Eder
Institute for Process and Particle Engineering Graz University of Technology, Graz, Austria

Michael Narodoslawsky
Institute for Process and Particle Engineering Graz University of Technology, Graz, Austria

Toshiyuki Hokari
Shimizu Corporation, Institute of Technology, Japan

Teruki Iwatsuki
Tono Geoscientific Research Unit Japan Atomic Energy Agency, Japan

Takanori Kunimaru
Tono Geoscientific Research Unit Japan Atomic Energy Agency, Japan

Geoffrey P. Hammond
Department of Mechanical Engineering, University of Bath, Bath., UK

Craig I. Jones
Circular Ecology Ltd., Bristol., UK

Rachel Spevack
Department of Mechanical Engineering, University of Bath, Bath., UK

Gabriel Anzaldi
Barcelona Digital Technology Center, Barcelona, Spain

www.ingramcontent.com/pod-product-compliance
Lightning Source LLC
Chambersburg PA
CBHW050455200326
41458CB00014B/5185